Isabel Stabile,
Tim Chard and
Gedis Grudzinskas

Clinical Obstetrics and Gynaecology

Second Edition

Isabel Stabile, MD, MRCOG
Center for Prevention and Early Intervention Policy, Florida State University, 1339 East Lafayette Street, Tallahassee, Florida FL 32301, USA

Tim Chard, MD, FRCOG
Reproductive Physiology Laboratory, Department of Obstetrics and Gynaecology, St Bartholomew's Hospital, West Smithfield, London EC1A 7BE, UK

Gedis Grudzinskas, MB BS, MD, FRCOG, FRACOG
Department of Obstetrics and Gynaecology, 4th Floor, Holland Wing, The Royal London Hospital, Whitechapel, London E1 1BB, UK

ISBN 1-85233-615-3 2nd edition Springer-Verlag London Berlin Heidelberg
ISBN 3-540-19942-X 1st edition Springer-Verlag Berlin Heidelberg New York

British Library Cataloguing in Publication Data
Stabile, Isabel, 1957–
 Clinical obstetrics and gynaecology. – 2nd ed.
 1.Obstetrics 2. Gynecology
 I.Title II.Chard, T. (Tim) III.Grudzinskas, J. G. (Jurgis Gediminas)
 618
ISBN 1852336153

Library of Congress Cataloging-in-Publication Data
Clinical obstetrics and gynaecology / [edited by] Isabel Stabile, Tim Chard, and Gedis Grudzinskas.– 2nd ed.
 p.; cm.
 Includes index.
 ISBN 1-85233-615-3 (alk. paper)
 1. Generative organs, Female–Diseases. 2. Pregnancy–Complications. I. Stabile, Isabel, 1957– II. Chard, T. III. Grudzinskas, J. G. (Jurgis Gediminas)
 [DNLM: 1. Pregnancy complications. 2. Genital Diseases, Female.
 WQ 240 C6397 2000]
 RG101 .C6814 2000
 618–dc21   99-056572

Apart from any fair dealing for the purposes of research or private study, or criticism or review, as permitted under the Copyright, Designs and Patents Act 1988, this publication may only be reproduced, stored or transmitted, in any form or by any means, with the prior permission in writing of the publishers, or in the case of reprographic reproduction in accordance with the terms of licences issued by the Copyright Licensing Agency. Enquiries concerning reproduction outside those terms should be sent to the publishers.

© Springer-Verlag London Limited 2000
Printed in Great Britain

First published 1996; Second edition 2000

The use of registered names, trademarks, etc. in this publication does not imply, even in the absence of a specific statement, that such names are exempt from the relevant laws and regulations and therefore free for general use.

Product liability: The publisher can give no guarantee for information about drug dosage and application thereof contained in this book. In every individual case the respective user must check its accuracy by consulting other pharmaceutical literature.

Typeset by Florence Production Ltd, Stoodleigh, Devon
Printed at the Athenæum Press Ltd, Gateshead, Tyne & Wear, UK
28/3830-543210 Printed on acid-free paper SPIN 10718883

Preface

The aim of this book is to provide a straightforward summary of the knowledge required for examinations in specialist Obstetrics and Gynaecology. Part Two of the examination for Membership of the Royal College of Obstetricians and Gynaecologists would be a good example. The volume is intended as a companion to the highly successful *Basic Sciences for Obstetrics and Gynaecology* which covers the knowledge required for preliminary examinations.

Increasingly, examinations of all types are based on multiple choice questions (MCQ) or structured answer questions (SAQ). No apology is made for the fact that the present book addresses the sort of "fact" which lends itself to testing by this approach. Thus, there is little discussion of speculative or contentious areas, no account of present or future research, and no references. Numerous excellent books are available which cover these topics in a much fuller and more discursive manner, and the present volume does not seek to emulate them.

Even the most apparently immutable facts are subject to periodic revision. We have attempted to present the "state-of-the-art": most of the material is generally if not universally accepted. A particular problem arises with numerical information. Frequencies of diseases, frequency of clinical findings, efficiency of diagnostic tests and therapies, have almost always been the subject of numerous different studies, each of which yields somewhat different results. Thus, with a few exceptions, most of the numbers given in this book have a "correction interval" of plus or minus 50%. For example, if a figure of 30% is given as the five-year survival of a given stage of a particular tumour, it is likely that any figure between 20% and 40% would be acceptable to examiners as a correct answer.

For most purposes we have assumed that the reader has a background knowledge equivalent to that of a junior to middle-grade trainee in Obstetrics and Gynaecology. No book of this type can possibly be encyclopaedic. We do not claim to identify every conceivable fact which might be asked in an examination nor do we provide detailed "recipes" for practical procedures. However, we believe that we have covered sufficient ground to enable someone of moderate retentive memory to obtain a passmark

Finally, we hope that the book will also provide an ongoing and useful source of quick factual reference for all those involved in clinical Obstetrics and Gynaecology.

London, January 2000

Isabel Stabile
Tim Chard
Gedis Grudzinskas

Contents

Section I OBSTETRICS

Chapter 1 Obstetric Statistics 3
Birth Rate 3
Maternal Mortality 4
Fetal, Neonatal and Postneonatal Mortality 5

**Chapter 2 Diagnosis of Pregnancy and Assessment of
Gestational Age** 7
Diagnosis of Pregnancy 7
Assessment of Gestational Age 7
Hyperemesis Gravidarum 9

Chapter 3 Antenatal Care 11
Tests of Fetal Well-being 12

Chapter 4 Miscarriage 15
Causes of Miscarriage 15
Threatened Miscarriage 17
Inevitable and Incomplete Miscarriage 17
Missed Miscarriage 17
Cervical Incompetence 18
Congenital Anomalies of the Uterus 18
Recurrent Miscarriage 19
Septic Miscarriage 19
Induced Abortion 20

Chapter 5 Ectopic Pregnancy 23
Causes of Ectopic Pregnancy 23
Clinical Features of Ectopic Pregnancy 24
Treatment of Ectopic Pregnancy 24
Abdominal Pregnancy 24

Chapter 6 Gestational Trophoblastic Tumours 27
Hydatidiform Mole 27
Choriocarcinoma 29

Chapter 7 Congenital Abnormalities ... 31
Genetic Defects ... 31
Environmentally Induced Abnormalities ... 34
Multisystem Abnormalities ... 34
Screening for Congenital Abnormalities ... 42

Chapter 8 Rhesus Disease ... 45
Diagnosis ... 45
Management ... 46
Prognosis ... 46

Chapter 9 Hypertensive Disorders of Pregnancy ... 47
Classification ... 47
Pregnancy-Induced Hypertension ... 48
Gestational Hypertension ... 48
Pre-eclampsia ... 48
Eclampsia ... 50
Management of Fluid Retention in Pregnancy ... 51
Prognosis of Hypertensive Disease in Pregnancy ... 51

Chapter 10 Antepartum Haemorrhage ... 53
Abruptio Placentae ... 53
Placenta Praevia ... 54
"Indeterminate" APH ... 55
Disseminated Intravascular Coagulation (DIC) ... 55

Chapter 11 Disorders Involving Amniotic Fluid ... 57
Hydramnios ... 57
Oligohydramnios ... 57
Amniotic Fluid Embolism ... 58
Intra-amniotic Infection ... 58

Chapter 12 Premature Labour ... 59
Management ... 59
Outcome ... 60
Perinatal Mortality and Morbidity ... 61
Preterm Premature Rupture of Membranes ... 61

Chapter 13 Multiple Pregnancy ... 63
Mechanisms of Twinning ... 63
Complications of Multiple Pregnancy ... 64
Antenatal Management of Twins ... 65
Management of Labour ... 65

Chapter 14 Maternal Diseases in Pregnancy ... 67
Cardiovascular Disease ... 67
Respiratory System ... 70
Neurological Disease ... 74
Genitourinary Disease ... 76
Musculoskeletal Disease ... 80

Gastrointestinal Disease 83
Endocrine Diseases 85
Acute Abdominal Pain in Pregnancy 93
Malignant Disease 95
Skin Disease 96
Haematological Disorders 99
Haematological Malignancies 105

Chapter 15 Normal Labour 107
Mechanisms and Course of Labour 107
Normal Delivery 108
Induction of Labour 109
Analgesia 109

Chapter 16 Abnormal Labour 111
Prolonged Labour 111
Abnormal Presentation 112
Monitoring of Fetal Well-being During Labour 115
Delay in Second Stage 117
Prolapse of the Cord 117
Ruptured Uterus 118
Impacted Shoulders 118

Chapter 17 The Puerperium 119
Postpartum Haemorrhage 119
Infection 120
Other Urinary Tract Problems 121
Psychiatric Disorders 121
Thromboembolism 122
Breast Feeding and Breast Problems 123

Chapter 18 The Neonate 125
Examination of the Newborn 125
Asphyxia 125
Resuscitation of the Newborn 127
Specific Problems of the Neonate 128

Chapter 19 Obstetric Operations 133
Amniocentesis 133
Chorionic Villus Sampling 133
Fetoscopy and Cordocentesis 134
Termination of Pregnancy 134
Sterilisation 134
Cervical Suture 135
Version for Breech Presentation 135
Caesarean Section 136
Forceps Delivery 136
Ventouse Delivery 137
Episiotomy and Repair (Including Tears) 137
Manual Removal of Placenta 138

Section II GYNAECOLOGY

Chapter 20 Menstrual Disorders 141
Puberty 141
Menorrhagia 141
Mechanisms of Menstruation 142
Primary Amenorrhoea 144
Secondary Amenorrhoea 146
Premature Menopause 147
Dysmenorrhoea 147
The Premenstrual Syndrome 147
Dyspareunia 148

Chapter 21 Menopause 149
Stages of the Climacteric 149
Postmenopausal Endocrinology 149
Anatomical Changes in the Climacteric 150
Pathology of the Climacteric 150
Clinical Features 151
Types of HRT and its Side-Effects 152
Postmenopausal Bleeding 152

Chapter 22 Virilism and Hirsutism 153
Virilism 153
Hirsutism 153

Chapter 23 Genital Infections 155
Infections and Related Conditions of the Vulva 155
Vaginal Infections 157
Cervicitis 159
Pelvic Infections 159
Other Infections 160
Chronic Pelvic Inflammatory Disease 160
Other Sexually Transmitted Diseases (Non-genital) 161

Chapter 24 Benign and Malignant Lesions of the Vulva 163
Benign Tumours of the Vulva 163
Other Benign Vulval Swellings 163
Vulval Dystrophies 164
Premalignant Conditions of the Vulva 164
Carcinoma of the Vulva 165
Rare Malignant Tumours of the Vulva 166

Chapter 25 Benign and Malignant Lesions of the Vagina 167
Benign Tumours of the Vagina 167
Carcinoma of the Vagina 167

Chapter 26 Benign and Malignant Lesions of the Cervix 169
Benign Tumours of the Cervix 169

Carcinoma-in-situ of the Cervix 169
Microinvasive Carcinoma 170
Carcinoma of the Cervix 171
Carcinoma of the Cervix in Pregnancy 173

Chapter 27 Benign and Malignant Lesions of the Endometrium 175
Benign Tumours of the Endometrium 175
Carcinoma of the Endometrium 175

Chapter 28 Benign and Malignant Tumours of the Myometrium 179
Benign Tumours of the Myometrium: Fibroids 179
Fibroids During Pregnancy 180
Metastasising Fibroids 180
Malignant Non-epithelial Tumours of the Uterus 180

Chapter 29 Tumours of the Ovary 183
Benign Tumours of the Ovary 183
Special Tumours of the Ovary 184
Malignant Epithelial Tumours of the Ovary 186

Chapter 30 Infertility 189
Aetiology 189
Management 189

Chapter 31 Contraception 193
Natural Family Planning 193
Barrier Methods 194
The Intrauterine Contraceptive Device 195
Hormonal Contraception 196

Chapter 32 Genitourinary Tract Disorders 199
Urodynamic Investigations 199
Stress Incontinence 200
Detrusor Instability 200
Retention of Urine 201
Urinary Tract Infections (UTIs) 201
Fistulae 201
Prolapse 202

Chapter 33 Endometriosis and Adenomyosis 203
Endometriosis 203
Adenomyosis 205

Chapter 34 Congenital Uterine and Vaginal Abnormalities ... 207

Chapter 35 Gynaecological Operations 209
Vulvectomy 209
Vaginal Repair 209

Colposcopy and Cervical Operations 210
Dilatation and Curettage 210
Hysterectomy 210
Tubal Surgery 211
Ovarian Surgery 211
Laparoscopy 212
Hysteroscopy 212
Pre- and Postoperative Care 212

Subject Index **213**

Section I
Obstetrics

Chapter 1

Obstetric Statistics

Birth Rate

A rate is a means of expressing a raw number as a proportion of another number. Since the result is a fraction with a value less than 1, it is multiplied by 1000. The crude birth rate relates the total number of births (numerator) to the total population (denominator). The birth rate reflects the impact of fertility on population growth and is expressed as follows:

$$\frac{\text{the number of live or dead births in a year} \times 1000}{\text{mid-year population}}$$

Underdeveloped countries typically have birth rates of 50 or more; developed countries have rates of 20 or less.

The general fertility rate relates the number of births to the population at risk, i.e. women in the childbearing ages (assumed to be 15–44 years). It is expressed as follows:

$$\frac{\text{the number of live births to women of all ages in a year} \times 1000}{\text{mid-year population of women 15–44 years old}}$$

The general fertility rate is often used to estimate the expected number of births in future years. African-American women in the USA have significantly higher fertility rates than white women.

In the UK there is a statutory requirement for parents to register a birth with a local Registrar within 42 days of delivery. The local Registrar reports to the Registar General. A midwife attending a birth must inform the local Medical Officer within 36 hours of the birth. All 50 states in the USA have statutes requiring that a birth certificate be completed for every birth and submitted promptly to the local Registrar.

Maternal Mortality

In the UK, maternal mortality is defined as death attributable to pregnancy occurring during the pregnancy or within 6 weeks after delivery or abortion. The maternal mortality rate is:

$$\frac{\text{the number of maternal deaths} \times 1000}{\text{total births}}$$

Factors that may confuse these numbers are: (a) some deaths are fortuitous (e.g. road traffic accidents) and may or may not be included; (b) total births do not include abortions because the latter are not registered. In the UK the maternal mortality rate has fallen from 4 in 1000 in the 1930s to less than 0.1 in 1000 in the 1990s.

Maternal deaths in the UK are reported to the Registrar. The detailed report is reviewed by a senior Regional obstetrician, together with others as appropriate (pathologist, anaesthetist, etc.). The Confidential Enquiry into Maternal Deaths, which is published triennially, analyses details of virtually every maternal death in the UK and classifies them into: (1) indirect obstetric deaths, those resulting from pre-existing disease or disease that developed during pregnancy that was aggravated by pregnancy; (2) fortuitous deaths, due to causes which just happened to occur during pregnancy; or (3) true (or direct) deaths resulting from obstetric complications of pregnancy, labour and puerperium, from interventions, omissions, incorrect treatment or from a chain of events resulting from any of the above. The major direct causes of maternal death are pulmonary embolism, toxaemia, haemorrhage, ectopic pregnancy, miscarriage, sepsis and amniotic fluid embolism.

In the triennium 1994–6, the major causes of maternal death per million maternities in the UK were thromboembolism (21.8%), hypertensive disorders (9.2%), amniotic fluid embolism (7.7%), early pregnancy complications (6.8%), sepsis (6.4%) and haemorrhage (5.5%).

1. *Thrombosis and thromboembolism*: two thirds of cases occur after delivery, often without warning. The origin is usually the pelvic or leg veins. The risk is increased with operative delivery, obesity and advanced maternal age. Avoidance includes early diagnosis of leg vein thrombosis and early use of anticoagulation. If there has been a previous thrombosis some advise anticoagulation around the time of delivery.
2. *Haemorrhage*: postpartum haemorrhage is the commonest cause of death from haemorrhage, followed by abruptio placentae and placenta praevia. Deaths from all types increase with age and parity.
3. *Hypertensive disorders*: This includes hypertensive disorders of pregnancy, eclampsia and liver disease.

A significant number of maternal deaths are associated with anaesthesia; 50% of these are due to aspiration of stomach contents leading to Mendelson's syndrome. The experience of the anaesthetist is an important determining factor, as is prevention with antacids (sodium citrate, ranitidine, etc.).

A woman's death is classified as pregnancy related in the USA if it occurs during pregnancy or within one year after the pregnancy. From 1987 to 1990, pregnancy-related maternal mortality increased from 7.2 to 10 in 100 000 live births, probably reflecting improved surveillance and reporting systems. The Healthy People 2010 goal is no more than 3.3 maternal deaths per 100 000 live births overall. The rate is four times greater for African-Americans than for whites. The leading causes of pregnancy-related maternal death in the USA are thromboembolism, haemorrhage and hypertensive disorders.

Fetal, Neonatal and Postneonatal Mortality

In the UK, a stillbirth is a baby born after 24 weeks' gestation that shows no signs of life after separation from the mother. A neonatal death is a baby born alive who dies within 28 completed days of birth. A perinatal death may be a stillbirth or first week neonatal death. A fetus born without signs of life before 24 weeks is a miscarriage, though if there are brief signs of life it becomes a neonatal death. International comparisons are hampered by differing reporting requirements. In the USA, fetal deaths are those that occur prior to the complete expulsion from the mother, irrespective of the duration of pregnancy. Reporting requirements (i.e. specific gestational age and weight limits) for fetal deaths vary from state to state.

The perinatal mortality rate (PNMR) is:

$$\frac{\text{stillbirths} + \text{1st week neonatal deaths} \times 1000}{\text{total births}}$$

The PNMR is higher in the USA than in the UK (partly due to reporting differences), but lower in many Scandinavian and European countries, including France. Unlike postneonatal mortality, stillbirths and neonatal mortality rates have changed very little between 1993 and 1995 in the UK. The PNMR is lowest for mothers aged 20–30 years in their second pregnancy. There is a direct inverse relationship to socioeconomic class and a close relation to birthweight. The single commonest cause of perinatal death in the UK is congenital abnormality. Other common causes are low birthweight (less than 2500 g), prematurity, asphyxia, birth injury, haemolytic disease, respiratory distress syndrome and infections.

A variety of factors have contributed to the fall in PNMR, including better general health of the population, better obstetric care and the widespread introduction of neonatal intensive care, leading to better survival of low birthweight babies. In the more recent past there has been a 50% reduction in deaths from congenital malformation due to improved prenatal diagnosis, preregistration abortion and improved survival after treatment. However, there has been no significant change in the rate of death from the consequences of prematurity, in spite of gradually improving birthweight-specific survival. The perinatal mortality rate may now be an inappropriate indicator of standards of obstetric care because it is uncommon, more than half the deaths are unexplained and the two major causes – congenital abnormalities and preterm delivery – are as yet largely unpreventable. Thus, since 1993, in England, Wales and Northern Ireland, obstetric practice has been monitored by the Confidential Enquiry into Stillbirths and Deaths in Infancy (CESDI) covering all deaths from the twentieth week of pregnancy till the end of the first year of life. Confidential assessment by multidisciplinary panels of specialists at Regional level provide data for a national level annual report that highlights factors contributing to deaths

Almost 80% of intrapartum-related deaths in the most recent report available (1994–5) were criticised for suboptimal care.

Postneonatal deaths are those that occur to liveborns after 28 days but before the first year of life. The death rate from infection, sudden infant death syndrome (SIDS) and congenital malformations in the postneonatal period continues to fall in the UK.

The combination of neonatal and postneonatal deaths equals the infant mortality rate, which is expressed as follows:

$$\frac{\text{infant deaths} \times 1000}{\text{live births}}$$

All states within the USA require that all infant deaths be reported, though the birthweight and gestational age are not recorded on the death certificate. Most states link birth and death certificates to report weight- and gestation-specific mortality rates. Infant mortality is more than double in African-Americans than in white mothers in the USA.

Chapter 2
Diagnosis of Pregnancy and Assessment of Gestational Age

Diagnosis of Pregnancy

The symptoms of early pregnancy are: (a) amenorrhoea (but some irregular bleeding may occur from the decidua parietalis); (b) nausea and vomiting; (c) breast enlargement, with tingling and tenderness (from 6 weeks); (d) darkening of the areola (from 8 weeks), which persists after pregnancy; (e) frequency of micturition; and (f) other symptoms, including tiredness (the most common), excess salivation, pica and mild irritability or euphoria.

The signs of early pregnancy are: (a) breasts are swollen and warm, with prominent superficial veins and darkening of the areolae; (b) uterus and cervix feel soft and enlarged; (c) cervix appears bluish rather than pink.

The standard pregnancy test is a qualitative measurement of chorionic gonadotrophin (hCG) in urine. This uses an antibody to the beta subunit of hCG, and there is therefore no interference from luteinising hormone (LH). Current tests with a sensitivity of 50 units/litre or better will become positive in most women at around the time of the missed period, i.e. 7 days after implantation. Most over-the-counter pregnancy tests are based on enzyme-linked immunosorbent assays that give a colorimetric indication of hCG levels within about 5 min. Very sensitive hCG tests may give positive results for many weeks after a pregnancy has been delivered, failed or terminated. A gestational sac can be detected by transvaginal ultrasound (TVS) a few days after the missed period and 1–2 weeks later with transabdominal ultrasound (TAS). Fetal tissues can be seen from 6 weeks onwards and a fetal heart from 7 weeks using TAS and 1 week earlier using TVS. Portable machines (Sonicaid, Doptone) that measure blood flow based on the Doppler principle can be used to detect the fetal heart beat from the tenth week onwards.

Assessment of Gestational Age

Traditionally, estimation of gestational age is based on the menstrual history and the findings of bimanual and abdominal examination. Both of these may be subject to considerable error. Menstrual dating is particularly unreliable in patients with irregular menstrual cycles and those using hormonal contraception prior to conception. Many maternity units in the UK now perform ultrasound measurements as a routine in all pregnancies to provide an objective estimate of gestational age. In the USA and Canada, universal ultrasound screening of all pregnancies is recommended only in the presence of recognisable risk factors.

Ultrasound energy delivered to tissues varies with: (a) frequency (transvaginal transducers use higher frequencies than transabdominal ones); (b) intensity (power); (c) exposure time (24 h of scanning with pulsed ultrasound produces only about 80 s of ultrasound exposure); and (d) distance from emitting source (since energy delivered to tissues varies inversely with the square of the distance from the source, even when the transducer is only a few centimetres from the target (e.g. TVS), the fetus receives less than 0.1% of the maximum safe levels). There are no known examples of damage to target tissues from the use of conventional diagnostic ultrasound imaging.

A variety of ultrasound parameters have been proposed for assessment of gestational age. All are subject to some observer error. The accuracy of gestational age estimates by ultrasound increases as more variables are measured and is inversely related to fetal age. Later in pregnancy the accuracy is enhanced by serial measurements. Other significant variables include sex (males are larger than females after 30 weeks), race (Europeans > Asians > Africans), socioeconomic status (class I children are larger than class V), multiple pregnancy, malformations, etc. The estimate of fetal age is determined from nomograms derived from a variety of study populations. The most commonly used measurements are the following.

Crown-Rump Length (CRL)

This is the distance between the crown of the head and the caudal end of the torso. Careful sonographers will take the measurement three times. The results should be within 1 mm of each other. Between 7 and 12 weeks' gestation this gives an estimate of gestation accurate to within ±3.5 days. After 10 weeks the accuracy decreases as the fetus flexes and becomes more difficult to measure.

Biparietal Diameter (BPD)

This is the diameter of an oval plane in which the cavum septum pellucidum is seen anteriorly and the thalamic nuclei posteriorly. The BPD is measured from the outer calvarium on one side to the inner calvarium on the opposite side, on a line running just anterior to the thalamic nuclei. Between 14 and 20 weeks' gestation this gives an estimate of gestation accurate to within ±1 week. In late gestation the error is 3–4 weeks.

Other Ultrasound Parameters of Gestational Age

1. Gestational sac diameter (4–6 weeks).
2. Head circumference; this is similar, but not superior, to BPD.
3. Femur length (14–20 weeks); gives results similar to BPD.

4. Fetal abdominal circumference measured at the level of the intra-abdominal portion of the umbilical vein; it is accurate to within ±12 days prior to 20 weeks, and ±21 days after 36 weeks; it is also used for estimation of fetal weight.
5. Ancillary measures such as distal femoral epiphyseal size, intraorbital diameter and renal diameter can be used.

Hyperemesis Gravidarum

Hyperemesis is a rare complication of the first trimester (3.5 per 1000 pregnancies). It is commoner in hydatidiform mole (25% of cases) and multiple pregnancy, and an association with high levels of hCG has been suggested. In 90% of cases the corpus luteum is in the right ovary. It may be the first sign of thyrotoxicosis in pregnancy. Starvation and fluid and electrolyte disturbances may occur. Management includes small frequent meals; if more severe, parenteral fluid, electrolytes and antiemetics are given.

Chapter 3

Antenatal Care

The primary aim of antenatal care is to predict, and thereby hopefully to prevent, one or more of the untoward outcomes listed in Table 3.1.

Classically, a pregnant woman is seen monthly until the 28th week, then every 2 weeks until the 36th week, then weekly until delivered. Many believe that in a woman with no risk factors these intervals can be safely extended. The standard observations made at any antenatal visit are listed in Table 3.2. More specific tests of fetal growth and well-being are considered below.

Table 3.1. Unsatisfactory fetal outcomes that might be predicted by clinical observations in a pregnant woman

1. Perinatal death, stillbirth or neonatal death
2. Intrauterine growth retardation
3. Fetal distress in labour
4. Neonatal asphyxia
5. Postnatal motor and intellectual impairment
6. Preterm delivery
7. Congenital abnormalities
8. Specific fetal diseases, such as erythroblastosis and the metabolic and nutritional disorders induced by maternal diabetes

Table 3.2. Routine examinations at every visit

Maternal weight
Urinalysis
Blood pressure
Check for oedema
Measure symphysis-fundal height

Tests of Fetal Well-being

Numerous special tests of fetal well-being have been advocated. Few are universally accepted in clinical practice. The problems of assessment and implementation include: (a) the occurrence of false-negative results; (b) the occurrence of false-positive results, which are more serious because they can dictate actions (induction, caesarean section) that themselves carry a risk; and (c) the lack of any proven form of therapy for the "at-risk" fetus.

Maternal Weight Gain

From the 20th to the 36th week average weight gain should be between 250 and 650 g per week. There is no association between fetal risk and weight gain, though an initial weight of less than 45 kg or more than 100 kg does constitute a risk.

Fetal Movements

The mother should feel fetal movements on average once every 10 min, and at a rather lower frequency near term. Reduced fetal movements (less than three movements per 30 min on average, or less than four episodes per 12 h in severe cases) are believed to be associated with unsatisfactory fetal outcome. The clinical value of this test has been questioned, though there is no doubt that absence of fetal movement is associated with an increased perinatal mortality rate.

Placental Function Tests

This was the name given to measurement of biochemical products of the fetus and placenta in the maternal circulation. The commonest tests were measurement of placental lactogen in maternal blood, and oestriol in maternal blood or urine. Both tests had a sensitivity of 30–50% in respect of unsatisfactory fetal outcome (Table 3.1), in particular for fetal growth retardation. These tests have now been largely replaced by biophysical procedures.

Ultrasound Measurement of Fetal Growth

Estimates of fetal weight in utero (and, thereby, prediction of birthweight) can be derived from measurement of the biparietal diameter (BPD), transabdominal circumference (TAC), femur length, or combinations of these. The error of each of these estimates is ±16% (2 standard deviations (SD)) provided that the stage of gestation is accurately known. Estimates of the sensitivity of ultrasound in respect of growth retardation (birthweight less than the 10th centile) range from 40% to 80%; there is also a high incidence of false-positive results.

It has been claimed that serial measurement of BPD can reveal two types of growth retardation: that in which there is a sudden deviation from a previously normal pattern; and that in which growth is parallel to, but at a lower level than, the normal range. The clinical value of this distinction is doubtful. It is also believed that growth problems arising before 24 weeks lead to symmetric growth retardation, while those that arise after 24 weeks lead to asymmetric growth retardation. Because the liver grows faster than the head during the last weeks of pregnancy, TAC might be a better index of fetal nutrition at that time than the BPD. Combination of TAC and BPD is said to be useful in the diagnosis of asymmetric growth retardation.

Antepartum Cardiotocography (CTG)

A continuous recording is made of the fetal heart rate (FHR) and fetal movements for 30 min or longer. The characteristics of a normal FHR are:

1. Baseline rate 110–160 beats per minute (bpm).
2. Beat-to-beat variation of 10–25 bpm. Loss of variability is associated with fetal hypoxia, infections and administration of central depressant drugs. Variability increases when the fetus is awake and in association with fetal breathing. A "sinusoidal" pattern of FHR is associated with fetal anaemia.
3. Accelerations: transient accelerations (15–20 bpm) are found in association with fetal movements, external stimuli and uterine contractions. In a normal "reactive" trace there should be two or more FHR accelerations, by 10–15 bpm for 15 s, in a period of 30 min.
4. Decelerations should be absent or rare.

The trace may be interpreted by simple inspection, or by one of a variety of scoring systems that seek to combine various features including variability and accelerations. A normal "reactive" trace is considered to be a good index of fetal health, even if some decelerations are also seen. A "non-reactive" trace suggests that the fetus is at risk. However, since prolonged periods of inactivity can occur in a normal fetus for 75 min, the test time can be extended or attempts can be made to stimulate the fetus. A "non-reactive" trace for 120 min has a predictive value of 80% or more for fetal mortality and morbidity. The best use of antepartum CTG is as a serial test in high risk pregnancy. Some have advocated the use of a "stress" as part of the test (oxytocin challenge, nipple stimulation, acoustic or light stimulation, exercise) but stress tests are generally associated with high false-positive rates and are of no proven benefit.

Amniotic Fluid Volume

Ultrasound should reveal at least one pool of amniotic fluid measuring at least 2 cm in two planes. Oligohydramnios suggests abnormalities of the fetal kidneys, growth retardation or premature rupture of the membranes.

Fetal Breathing Movements (FBM)

In the normal fetus there may be at least 30 s of FBM in every 30 min, although sometimes the interval between episodes of FBM may be an hour or more. Absent or reduced FBM may occur in fetal hypoxia and growth retardation. The test is little used clinically.

Biophysical Profile

This is a combination of measurements of amniotic fluid volume, fetal movements, fetal breathing movements, fetal tone and fetal reactivity. The value of this test in routine as opposed to research practice has yet to be proved.

Doppler Blood Flow

The uteroplacental and umbilical circulations are normally "low resistance", i.e. there is always forward flow even at the end of diastole. Failure to create a low resistance uterine circulation by 20 weeks is revealed by bilateral "notching" of the uterine arteries; this is associated with an increased risk of developing pre-eclampsia or growth retardation. Loss of end-diastolic flow in the umbilical arteries suggests that the outlook for the fetus is unsatisfactory. This appears to be a useful test to confirm or exclude fetal jeopardy in cases in which the fetus is already thought to be at high risk. It is not a useful screening test in an unselected population.

Chapter 4

Miscarriage

Miscarriage is variously defined as: the expulsion or extraction of a fetus (embryo) weighing less than 500 g equivalent to approximately 22 weeks' gestation (World Health Organization, 1977); or termination before fetal "viability" is achieved; or termination prior to 24 completed weeks of gestation (UK legal definition). The nomenclature of the various types of miscarriage is shown in Table 4.1. The incidence is 15% of clinically apparent pregnancies, but may be considerably higher if "occult" pregnancies are taken into account. Some 25% of women will have one or more miscarriages.

Causes of Miscarriage

1. *Fetal abnormality*: 40% or more of miscarried fetuses are abnormal (structural, chromosomal or genetic). The incidence of chromosomal abnormalities has been estimated at 30–60%. The commonest chromosome abnormality is trisomy (16, 22, 21, 15), followed by monosomy (usually 45,X), triploidy and

Table 4.1. Definitions of the various types of miscarriage. Note that in the first trimester bleeding is likely to come from the decidual space and that the fetus is often delivered before the placenta. In the second trimester bleeding is usually from the placental site, and the fetus and placenta are expelled together

Types of miscarriage	Definition/Description
Threatened	Bleeding from the uterus prior to 24 weeks with the cervix not dilated and the fetus alive
Inevitable	Bleeding from the uterus prior to 24 weeks with pain and dilatation of the cervix
Incomplete	Part of the conceptus has been expelled but there is persistent bleeding due to chorionic tissue remaining in the uterus
Complete	The whole conceptus has been expelled
Recurrent (habitual)	Three or more consecutive miscarriages
Missed	Pregnancy failure is identified before expulsion of fetal/placental tissues

tetraploidy. Most autosomal trisomies are secondary to non-disjunction during the first meiotic division of the oocyte, and unlike other chromosomally abnormal miscarriages are commoner with increasing maternal age. Structural abnormalities, including neural tube defects, are also associated with an increased incidence of miscarriage.

2. *Abnormalities of implantation*: these occur with an intrauterine device. Low implantation of the placenta may be a cause of mid-trimester miscarriage.
3. *Multiple pregnancy* secondary to fetal abnormalities.
4. *Intrauterine adhesions*: typically following overzealous postpartum curettage, endometritis or intrauterine surgery.
5. *Endocrine deficiencies*: early recurrent miscarriages have been attributed to a luteal phase defect. Though the corpus luteum is considered to be essential during the first 8 weeks of pregnancy, there is no evidence that primary failure of the corpus luteum after spontaneous ovulation is a specific cause of miscarriage.
6. *Uterine abnormalities* (congenital and diethylstilboestrol (DES)-induced uterine abnormalities, submucous fibroids): fusion abnormalities of the uterus may double the rate of mid-trimester miscarriage. The most severe abnormalities may carry a lower risk. Retroversion is not a cause of miscarriage except in the extremely rare cases in which a retroverted gravid uterus is trapped in the pelvis. Fibroids may cause miscarriage by interfering with implantation, or the hormonally induced increase in size might lead to mid-trimester loss by premature uterine contractions.
7. *Maternal disease*: miscarriage has been attributed to severe maternal illness, especially pyrexial infections, and to poorly controlled diabetes mellitus, thyroid disease, systemic lupus erythematosus and von Willebrand's disease. However, the only disorder clearly and specifically associated with miscarriage is Wilson's disease (an inherited disturbance of copper metabolism). Maternal age per se (over 35 years) leads to an increased risk of miscarriage.
8. *Infections*: fetal death and, therefore, miscarriage may be caused by microorganisms including treponemas (syphilis), rubella, variola, vaccinia virus (cowpox), polio virus, parvovirus (erythema infectiosum or fifth disease), herpes simplex, toxoplasma, cytomegalovirus, mycobacterium tuberculosis, trypanosoma (Chagas' disease), plasmodium (malaria), listeria, brucella, mycoplasma, ureaplasma, salmonella and vibrio.
9. *Poisons*: cytotoxic drugs can kill the fetus, as can high levels of lead, quinine, aniline, benzene, formaldehyde and, possibly, anaesthetic gases (higher incidence in operating theatre staff). Smoking and alcohol may have a slight association with miscarriage. Exposure to video display terminals does not cause miscarriage.
10. *Immunological factors*: rhesus incompatibility will occasionally cause a mid-trimester miscarriage. It has been proposed that a close similarity of histocompatibility antigens between the parents is a cause of recurrent miscarriage, but the evidence for this is inconclusive.
11. *Autoimmune disease*: lupus anticoagulant and anticardiolipin antibodies are associated with an increased likelihood of fetal wastage, largely in the second trimester.
12. *Cervical incompetence*: this may occur without predisposing factors, but more commonly is iatrogenic – the result of overforceful dilatation of the cervix

(more than 10 mm), obstetric injuries, amputation, cauterisation or cone biopsy. Elective first trimester termination is not now believed to be a cause of subsequent mid-trimester miscarriage. Cervical incompetence may also follow exposure to DES in utero.
13. *Trauma*: including amniocentesis and pelvic surgery.

Threatened Miscarriage (see Table 4.1)

Some sort of minor bleeding occurs in the early stages of 20–25% of all pregnancies. A single ultrasound examination is generally sufficient to differentiate a viable pregnancy from an incomplete or complete miscarriage, a missed miscarriage, a hydatidiform mole or an anembryonic pregnancy.

Approximately 50% of pregnancies complicated by threatened miscarriage have a successful outcome but if the gestational sac and its contents appear normal by ultrasound, and if a fetal heartbeat can be demonstrated (after 8 weeks) there is a 90% or better chance that the outcome will be satisfactory. Prognosis can also be determined by serial measurements of placental products such as hCG.

There is no specific treatment for threatened miscarriage. Once the pregnancy is confirmed as viable by ultrasound, rest (not necessarily in bed) is recommended. The possible benefits of progesterone therapy have not been confirmed. If anything, this may merely prolong the miscarriage process (i.e. produce a "missed" miscarriage). If an intrauterine device is present and the tail is visible, it should be removed.

In those pregnancies that continue there is an increased incidence of later problems, including premature labour, low birthweight and perinatal death.

Inevitable and Incomplete Miscarriage (see Table 4.1)

Once the diagnosis is made the uterus should be evacuated. The miscarriage process can be expedited and bleeding controlled with a combination of ergometrine (0.5 mg) and oxytocin (5 units) in the first trimester and prostaglandins in the second trimester. In the first trimester the process is often incomplete and evacuation by curettage and ovum forceps or suction is usually performed; in the second trimester this may not be necessary if ultrasound shows an empty uterine cavity. Supportive therapy (correction of blood loss, analgesia, antibiotics for infection) is given as necessary. If the mother is rhesus negative, anti-D immunoglobulin (Ig) (250 units) should be given.

Missed Miscarriage (see Table 5.1)

Missed miscarriage (or embryonic demise) may be preceded by the signs of threatened miscarriage. The signs and symptoms of pregnancy regress; the uterus ceases to grow and may diminish in size. There is sometimes a brownish discharge. Levels of hCG fall. On ultrasound there is no fetal heart and the gestation sac may be collapsed. Ultrasonic distinction from a hydatidiform mole may be difficult. After 16 weeks there may be radiological evidence of fetal death, including collapse of the fetal skeleton. On occasion the whole chorion is surrounded by layers of blood clot (carneous mole).

The main complications are: (a) infection; (b) a major coagulation disorder similar to that which may accompany placental abruption (this is rare if the fetus is retained for less than 1 month, but occurs in 25–30% of cases thereafter); and (c) psychological distress to the mother. In anembryonic pregnancy ("blighted ovum") the sonographically empty gestational sac previously contained an embryo; this may be indistinguishable from embryonic demise.

In the first trimester suction evacuation is performed; in the second trimester uterine activity is stimulated by oxytocin infusion after pretreatment with mifepristone or vaginal prostaglandin E2 (PGE2). This should be given in the smallest possible volume via an infusion pump, to avoid water intoxication.

Cervical Incompetence

The incidence is thought to be 1–2% of pregnancies. Classically this presents with membrane rupture and relatively painless rapid labour in mid-trimester (usually after 16 weeks), though atypical forms are common. In the non-pregnant state the condition can be identified clinically by a gaping cervix, or by passage of dilators (6–8 mm or more is characteristic), ultrasonography or premenstrual hysterography.

Treatment is by insertion of a strip of unabsorbable material in the substance of the cervix (MacDonald suture) after ultrasound confirmation of fetal viability. The suture is placed as close as possible to the internal os, after upward dissection of the bladder. The optimum time is at 14 weeks of pregnancy, though some defer operation until regular inspection of the cervix shows bulging membranes. Severe herniation can often be corrected by gravity prior to insertion of the stitch. The use of antibiotics, progestational agents, and beta mimetics confers no advantage. The overall success of operative treatment is quoted at 70–80%, though there is much dispute as to its real efficacy (one preterm delivery prevented for every 20 sutures inserted).

The suture should be removed immediately if the membranes rupture or there are expulsive uterine contractions. In the absence of this, some advocate removal at 38 weeks, and others a caesarean section with the suture in situ.

Congenital Anomalies of the Uterus

These are usually anomalies of fusion of the müllerian ducts, ranging from a complete uterus didelphys to a small septum at the fundus. They affect 1% of the female population. They are a cause of recurrent second trimester miscarriage, successive pregnancies tending to be longer. The diagnosis may be made at the time of evacuation but is frequently overlooked. It can be confirmed by hysterography or hysteroscopy. Associated renal tract abnormalities are excluded by intravenous pyelography.

Treatment is usually conservative, though after three or more unsuccessful pregnancies some sort of plastic operation may be considered (usually removal of a septum and reconstruction of the uterus: by hysteroscopy or the Strassman procedure). However, the latter leads to a seriously scarred uterus, with a potential risk of uterine rupture in subsequent pregnancies. About 30% of treated patients are infertile.

Recurrent Miscarriage (see Table 4.1)

Three or more consecutive miscarriages occur in 0.4–0.8% of pregnant women. The risk of further miscarriage after three consecutive miscarriages has been estimated as 30–70%. It is most commonly a chance phenomenon, but specific causes might include any of the factors listed above. The earlier concept of endocrine factors as a cause, and diagnostic and therapeutic measures that stemmed from this, are now entirely discredited. Occasional cases may be due to a balanced translocation in one of the parents. Immunological factors and immunotherapy (immunising the mother against paternal antigens) are currently the topic of much investigation and argument. This apart, treatment is only possible if a specific cause has been identified. Cases associated with lupus anticoagulant have been successfully treated with corticosteroids and/or aspirin.

Septic Miscarriage

Infection may occur with missed miscarriage and with incomplete miscarriage – especially that resulting from inexpert mechanical interference or from inadequate surgical evacuation in the first trimester. The history of preceding criminal abortion may be withheld, though evidence of lower genital tract injury is suggestive. The commonest organisms are *Escherichia coli*, streptococci (haemolytic, non-haemolytic, and anaerobic), *Staphylococcus aureus*; rare organisms include *Clostridia welchii*, *Cl. tetani* and *Cl. perfringens*. Though the infection is usually confined to the uterine cavity it may spread to other pelvic organs and to the general circulation. The clinical signs are those of infection and a miscarriage process, associated with an offensive vaginal discharge and lower abdominal pain. The cervix may remain closed.

Vaginal and cervical swabs, and blood and urine cultures, are taken for aerobic and anaerobic bacteriology. The treatment is antibiotic therapy (intramuscular ampicillin or cephalosporin; oral metronidazole and/or tetracycline). In the more severe cases intravenous therapy should be considered, including chloramphenicol and gentamicin. The uterus should be evacuated as soon as possible after antibiotic therapy has commenced.

A rare but important complication is endotoxic shock. Toxins (lipoprotein-carbohydrate complexes) released into the bloodstream from Gram-negative organisms, such as *E. coli*, *Klebsiella*, *Proteus*, *Pseudomonas* or *Bacteroides* can directly affect small blood vessels to cause circulatory collapse, and can also cause disseminated intravascular coagulation with microthrombi in the kidneys, liver and lungs, and a general coagulation deficiency. The patient is transferred to an intensive care unit and treatment includes immediate intravenous injections of penicillin, gentamicin and metronidazole. Large doses of hydrocortisone may be given (10 g over 24 h); the use of both vasoconstrictor (during the hypotensive phase) and vasodilator (to assist perfusion) drugs, as well as digoxin and diuretics has been advocated. Intravenous fluids (including blood if necessary) are given, the central venous pressure being the main guide to either under- or overtreatment;

urine flow should be maintained at 30–60 ml/h. Metabolic acidosis can be corrected with bicarbonate. Severe renal damage (renal cortical necrosis) due to microthrombi may necessitate dialysis. The uterus is evacuated only after the patient's general condition has been stabilised.

Shock may also result from the exotoxins of staphylococci, streptococci and *Cl. welchii* and *Cl. perfringens*. Circulatory collapse is a late event. The clostridial infections are particularly likely to produce haemolysis and kidney damage.

Induced Abortion

In Britain 1 in every 5 pregnancies is terminated: fewer than 2% are for fetal anomalies. Induced abortion may be legal or illegal. In the UK the 1990 Human Fertilization and Embryology Bill states that abortion can be performed because of risk to the physical or mental health of the pregnant woman or her existing children; the law now requires that the pregnancy should not exceed 24 weeks. This limit does not apply when there is a risk of grave permanent injury or death to the mother, or a substantial risk of serious handicap in the child. The procedure must be performed in a national Health Service (NHS) hospital or other approved place, and must be notified to the Chief Medical Officer on form HSA3 within 7 days.

In the first trimester the methods for inducing abortion are:

1. Cervical dilatation and curettage, supplemented with sponge and/or ovum forceps.
2. Suction evacuation (should be avoided at gestations of 6 weeks or less).
3. Hysterectomy.
4. Prostaglandins.
5. Progesterone antagonists (mifepristone; followed 36–48 h later by prostaglandins (Gemprost or misoprostol) in pregnancies of less than 8 weeks' gestation). In 5–10% of cases surgical evacuation is still needed either for heavy bleeding or because of failure of the combination treatment to induce abortion.

The pervaginal procedures may be carried out under local (cervical and paracervical) or general anaesthetic. Suction evacuation may be safer under local anaesthesia.

In the second trimester the methods are:

1. Prostaglandins (e.g. Gemprost), administered vaginally; 36 to 48 h after mifepristone orally, a toxic agent, e.g. urea, may be injected into the amniotic cavity to induce fetal death.
2. Intra-amniotic injection of hypertonic glucose or saline (now little used).
3. Hysterotomy.
4. Hysterectomy.
5. Progressive cervical dilatation by hygroscopic agents (laminaria tents), followed by surgical evacuation.

The risks of induced abortion increase with increasing gestation. The immediate risks of the procedure are: (a) trauma to the cervix, uterus or surrounding organs; (b) infection (the risk is halved by giving antibiotics prophylactically); and (c) haemorrhage. Later risks may arise from the long-term effects of pelvic sepsis and from damage to the cervix or intrauterine adhesions (Asherman's syndrome). The risk of death from abortion was less than 0.1 per 100 000 in 1994–6.

Since ovulation may occur as early as 2 weeks after abortion, contraception should begin as soon as possible. Making sex education and contraception freely available, particularly to young and single women, may reduce the numbers of unwanted pregnancies and hence abortions.

Chapter 5

Ectopic Pregnancy

Ectopic pregnancy was responsible for 5.5% of maternal mortality in the UK (1994-6) and is the major cause of maternal mortality in the first trimester in the Western world. It may occur in the tube (95%), the uterus (intramural, angular, cervical or in a rudimentary horn), the ovary, the broad ligament or elsewhere in the peritoneal cavity. The extratubal sites are usually secondary to extrusion from the tube. In the tube, the commonest site is the ampulla, followed by the isthmus. In the ampulla the conceptus is often extruded, whereas in the isthmus the tube is usually ruptured. Rupture of an ectopic in the interstitial part of the tube, though rare, is associated with particularly severe haemorrhage. Heterotopic pregnancy, a combined intrauterine and extrauterine pregnancy occurs rarely in spontaneous conceptions, but in up to 3% of clinical pregnancies after in vitro fertilisation (IVF).

In the Western world the rate of ectopic pregnancy is about 1% of all pregnancies, 20 per 1000 live births or 10-15 per 10 000 women ages 15 to 44. In the Caribbean it may be ten times higher. The recurrence rate is about 10%. The overall incidence has increased but the case fatality rate is falling. Morbidity and mortality are related to the delay between presentation and treatment.

Causes of Ectopic Pregnancy

The basic cause is delay to the passage of the fertilised ovum down the tube. The commonest pathology is damage to the ciliated epithelium and peristaltic activity of the tube due to chlamydial, gonococcal, tuberculous and other infections. The risk of ectopic pregnancy is increased in association with previous pelvic surgery, salpingitis and a previous ectopic pregnancy, as well as in women undergoing assisted conception and those exposed to DES in utero. In patients with an intrauterine device in situ there is a much increased frequency of tubal and ovarian pregnancy relative to intrauterine gestation.

The fertilised ovum implants in the tubal epithelium but almost immediately invades the muscular coat because there is no submucosa. There may be decidual reaction in the endometrium, but the timing of this is very variable.

Clinical Features of Ectopic Pregnancy

The common presenting features are lower abdominal pain, delayed or irregular menses, vaginal bleeding or brown discharge (due to shedding of the decidua) and syncope. With tubal rupture there is acute abdominal pain and clinical shock (pallor, fainting, hypotension and tachycardia); pain is typically referred to the shoulder tip or interscapular area due to irritation of the diaphragm by blood. Body temperature is normal and there may be a moderate leukocytosis. The uterus is slightly enlarged and a tender mass may be found to one side; vaginal examination should be avoided if there is a strong suspicion of an ectopic. Subacute presentations frequently give rise to diagnostic confusion which can usually be resolved by:

1. Rapid, sensitive hCG assay, which is usually positive in cases of abdominal pain due to pregnancy complications; if the serum hCG is rising, the likelihood that the pregnancy is extrauterine rather than intrauterine increases as the hCG doubling time increases. The half life of hCG can be used to distinguish miscarriage from ectopic pregnancy.
2. Ultrasound imaging to locate the site of pregnancy (up to a week earlier by TVS than TAS) and in less than 20% of cases, visualise the fetus outside uterine cavity. TVS can detect an intrauterine sac by 33 days equivalent to an hCG level of 1000 IU/l (first IRP), a yolk sac by 38 days and embryonic echoes with visible heart motion by 43 days from ovulation.
3. Laparoscopy, which reduces the rate of ruptured ectopic pregnancy.
4. Rarely, laparotomy. Culdocentesis and/or dilation and curettage (D&C) are rarely used nowadays.

Treatment of Ectopic Pregnancy

Immediate intravenous fluids are started with blood as necessary. In the most severe cases this may include heroic measures such as use of uncross-matched blood or even autotransfused blood. Without waiting for the outcome of resuscitative measures, a laparotomy is performed, followed by simple salpingectomy with conservation of the ovaries. If the tube has not ruptured then conservative surgery is recommended such as (laparoscopic) linear salpingostomy or partial salpingectomy. Fertility rates after salpingotomy via laparoscopy or laparotomy are comparable: about 50% will conceive again, 40% will have a live birth and 12% another ectopic pregnancy. Selected cases may be managed by laparoscopic or ultrasound-directed aspiration and/or local injection with drugs (potassium chloride, methotrexate or prostaglandins). There is a 5–10% risk of persistent trophoblast remaining after conservative treatment (both medical and surgical). In the absence of a history of infertility, the conception rate after medical treatment is 80%, with a 10% rate of recurrent ectopic pregnancy. If the patient is rhesus-negative, anti-D Ig should be given.

Abdominal Pregnancy

About 1% of all ectopics are abdominal, but they account for one fifth of all maternal deaths due to this condition. Abdominal pregnancy is almost always the result of secondary implantation of a primary tubal pregnancy. If the fetus dies and is retained it may become infected or calcified. There may be a persistent abnormal lie; fetal parts are readily palpated and the uterus (much enlarged) may be felt separately from the fetus. X-rays show maternal intestinal gas superimposed on the fetus. Ultrasound shows oligohydramnios and no clear outline of the gestation sac.

The fetus should be delivered as soon as viability is achieved, though perinatal loss is 75% or greater. In most cases the placenta should be left because of the risk of uncontrolled haemorrhage, though infection, adhesions, obstruction and coagulopathy may occur. The baby often shows evidence of pressure malformations.

Chapter 6

Gestational Trophoblastic Tumours

Two types of tumours are recognised:

1. *Hydatidiform mole*: oedematous and avascular villi and trophoblastic overgrowth. The classical "complete" mole has no fetus; in "partial" moles there are focal molar changes in the placenta and a fetus may be present. An invasive mole may show invasion of the myometrium and metastasis, which usually but not always regresses spontaneously.
2. *Choriocarcinoma*: large masses of anaplastic trophoblast invading muscle and blood vessels (but not fetal blood vessels). The villous pattern is generally lost. Vascular metastases in the lung and at the vaginal introitus are common.

Histological grading is often ambiguous and the key criterion is whether the disease is persistent or not, and whether it is metastatic or non-metastatic.

Hydatidiform Mole

Aetiology and Distribution

At least 95% of complete moles are female (46,XX), both X chromosomes being derived from the father. The haploid sperm duplicates its own chromosomes after meiosis. More rarely two sperms fertilise an empty egg leading to a 46,XY mole. The partial moles are usually triploid, two sperms having fertilised an ovum (69,XXX, XXY or XYY). Diploid and tetraploid partial moles have also been described. Partial moles were believed to be rare: in reality they are commoner than complete moles but the majority are spontaneously miscarried and therefore not recognised. As in normal cells, the mitochondrial DNA is of maternal origin. Women with molar pregnancies have a high incidence of balanced translocations.

The frequency ranges from 0.5 to 2.5 in 1000 pregnancies. The rates in Japanese are double those in white people. There is an increased risk in teenagers and women over 35, the rate rising 10-fold after age 40. Age, parity and gestational age at the time of diagnosis do not affect the risk of malignant sequelae. Women with a history of one mole have a 10-fold risk of recurrence.

The presentation is similar to that of a threatened miscarriage, but the size of the uterus is often excessive for the calculated gestation. Hyperemesis, large luteal cysts and early onset pre-eclampsia are common, probably associated with the excess mass of the trophoblast. Signs of thyrotoxicosis are apparent in occasional cases. In partial moles, many of these features may be absent.

The diagnosis is confirmed by ultrasound. In most cases there is no fetus and the vesical tissue has a characteristic "snowstorm" appearance. Rarely a fetus coexists with a mole (partial mole), confusing the typical ultrasound findings. Levels of hCG are elevated but often do not give a useful distinction between molar and normal pregnancy. A chest X-ray should be performed to exclude metastases.

Complications such as haemorrhage and sepsis are rare. The principal risk is choriocarcinoma, which occurs in 3% of cases; the risk is low with partial moles. A repeat molar pregnancy occurs in only 1% of subsequent pregnancies. Follow-up with an hCG assay is essential. The determination should be repeated every 1–2 weeks until hCG disappears, then monthly for 1 year and 3-monthly for a second year. Positive levels may persist for up to 6 months; if still present at 1 year then there is almost invariably choriocarcinoma present. Patients who have had a prior trophoblastic tumour of any type should have a further urine and serum hCG assay 3 weeks after each subsequent pregnancy. A high ratio of free beta subunit to intact hCG may also identify patients with a high risk of malignancy.

Treatment

Even when the uterus is very large, the treatment is suction curettage. Oxytocin infusion should not begin until a moderate amount of tissue has been removed to reduce the risk of embolisation of trophoblastic tissue to the lungs; the procedure should be completed with sharp curettage. Evacuation is repeated if bleeding persists or hCG levels are elevated after 6 weeks. In women who have finished childbearing, hysterectomy should be considered. Prophylactic chemotherapy for all cases is not currently favoured. High oestrogen pills are thought to be associated with an increased need for chemotherapy, so barrier contraception should be used until hCG levels are undetectable, after which hormonal birth control can safely be used. Another pregnancy may be attempted after 1 year of negative hCG titres.

Chemotherapy is indicated for: (a) hCG levels greater than 30 000 IU/l (urine) or 20 000 IU/l (serum) at 4–6 weeks post evacuation; (b) rising hCG levels (more than 50% over 2 weeks) or titres that plateau for 3 consecutive weeks, at any time after evacuation; (c) persistent uterine bleeding and positive hCG levels; (d) histological evidence of choriocarcinoma or the appearance of metastases. Diagnostic curettage is rarely helpful, because the malignancy is often deep in the myometrium and occasionally can result in uterine perforation, haemorrhage and the need for hysterectomy. Single-agent treatment with methotrexate (higher dose with folinic acid) or actinomycin D often results in complete remission in non-metastatic trophoblastic disease. Rarely, combination chemotherapy is required.

Choriocarcinoma

The frequency is one in 30 000 pregnancies in the West and it is three times commoner in the Far East. Fifty per cent are preceded by hydatidiform mole, 40% by normal pregnancy, 5% by abortion or ectopic pregnancy and 5% are of non-gestational origin. The greatest risk is in older women.

The clinical features are vaginal haemorrhage, abdominal or vaginal swelling, amenorrhoea and chest symptoms due to lung metastases (dyspnoea and haemoptysis). Intra-abdominal haemorrhage due to uterine perforation by tumour tissue may occur. Less common sites for metastases include the brain, the liver, the spleen, the kidneys and the buccal mucosa.

The diagnosis is confirmed by ultrasound, hCG determination, chest X-ray ("cannon-ball" or "snowstorm" appearance), and computed tomography (CT) of the chest and abdomen.

Treatment

The form of treatment is determined by whether or not the patient is judged to be at high or low risk based on the presence or absence of the following: brain and/or liver metastases, serum hCG levels greater than 40 000 IU/l, long duration (antecedent pregnancy more than 4 months before diagnosis), and failed chemotherapy.

For low risk cases (metastatic trophoblastic neoplasia with good prognosis) the treatment is chemotherapy with methotrexate (which inhibits the conversion of folic acid to folinic acid), or actinomycin D, until the hCG titre is normal. Other drugs are sometimes used in combination. Side effects include: (a) skin rashes; (b) gastrointestinal ulceration (stomatitis and proctitis); (c) leucocyte depression, to avoid infection the patient is given antibiotic cover and is barrier nursed under conditions of controlled asepsis (best achieved in specially designed units); (d) depression of erythropoiesis, transfusions are given as necessary; (e) alopecia, hair will subsequently regrow but a wig can be used as a temporary measure; (f) hepatocellular damage and jaundice; and (g) skin photosensitivity. Effects of overdosage can sometimes be reversed with autologous bone marrow transplants.

The progress of treatment is monitored by urine or blood hCG levels, chest X-rays, full blood counts, and liver function tests. In most cases, hCG becomes negative within 4 weeks. One additional course beyond the first negative hCG reading is required.

For high risk or recurrent cases a sequential combination of antitumour agents (etoposide, methotrexate, actinomycin D, vincristine and cyclophosphamide) is used. At least four courses of chemotherapy are given after the first negative hCG titre. Intracranial metastases can be treated with intrathecal methotrexate or cytosine arabinoside. Irradiation of non-resectable lesions is an option.

A delay of 12 months before conception is advised following cytotoxic chemotherapy to reduce the risk of teratogenesis and avoid false-positive hCG readings. Most of those desirous of pregnancy will conceive again and achieve a live birth. There is no increase in fetal abnormality in subsequent pregnancies. Many advise hysterectomy in women who do not desire subsequent pregnancies.

In those who do, hysterectomy is only indicated if there are severe complications such as haemorrhage, or if there is residual disease confined to the uterus. Solitary pulmonary metastases resistant to chemotherapy may be removed by thoracotomy.

Survival is now almost complete if the time lapse from the preceding pregnancy is less than 4 months (or 6 months from molar pregnancy) compared with 50% survival when the interval is 13–24 months. Patients who develop brain metastases while on chemotherapy have the worst prognosis.

Chapter 7
Congenital Abnormalities

Serious congenital abnormalities occur in 2% of live births, 30% of stillbirths and 18% of first week neonatal deaths, accounting for 25% of perinatal deaths. They may be due to (a) genetic (chromosomal or single-gene) defects, (b) environmental factors (alcohol, drugs, toxic chemicals), or (c) a combination (multisystem abnormalities).

Genetic Defects

These account for 25% of all congenital abnormalities.

Chromosomal Disorders

Chromosomes are analysed by examining the fetal karyotype during the metaphase of mitotic cell division. Anomalies of chromosome number (aneuploidy, usually by non-disjunction) or structure occur in 5% of first trimester and 1% of second trimester pregnancies, 0.6% of live births and 5% of stillbirths. Maternal but not paternal age is a risk factor.

Abnormalities of Chromosome Number

In polyploidy the entire chromosome set is duplicated: in triploidy there are an additional 23 chromosomes; in tetraploidy all 46 chromosomes are duplicated. One in 10 000 liveborns have triploidy and rarely survive; miscarriage or molar pregnancy usually occur. In trisomy and monosomy, there is an extra or an absent chromosome (Table 7.1).

Aneuploidy may be caused by (a) non-disjunction, failure of the chromatids to separate during metaphase of meiosis; (b) non-disjunction or anaphase lag during mitosis leading to mosaicism (c.f. chimaera resulting from the fusion of two separate zygotes); and (c) translocation, a portion of one chromosome is transferred to another. Translocation may be reciprocal, with exchange of material

Table 7.1. Major chromosomal disorders

Chromosome	Syndrome	Mechanism	Incidence	Prognosis
Trisomy 21	Down's	97% non-disjunction; 2% translocation; 1% mitotic errors	Overall 1 in 650 live births; >35 = 1 in 300, >40 = 1 in 100, >45 = 1 in 20; 70% of cases are born to women <35; reduced fetal products (AFP, UE3) and increased trophoblast products (hCG) increase the risk	50% of embryos miscarry; major cause of mental retardation; 10% recurrence if maternal balanced translocation; 2% recurrence if paternal; 1% recurrence after one affected child
Monosomy XO	Turner's	Non-disjunction in paternal meiosis unrelated to paternal age	150 in 10 000 conceptions; 1 in 10 000 girls at birth; declines with advancing maternal age	20% have congenital heart disease; no increased risk of mental retardation; infertility, hypertension and osteoporosis in adults
Trisomy 18	Patau's		>35 = 1 in 200; >40 = 1 in 60; >45 = 1 in 20	80% miscarry
Trisomy 13	Edward's		>35 = 1 in 200; >40 = 1 in 60; >45 = 1 in 20	80% miscarry

AFP, alphafetoprotein; uE3, unconjugated oestriol; hCG, human chorionic gonadotrophin.

between two non-homologous chromosomes (balanced). Unbalanced progeny have 46 chromosomes, some genes in single copies and others in triple copies, and abnormal phenotype. Translocation may also be robertsonian, with loss of short arms from each of two acrocentric chromosomes (13, 14, 15, 21, 22), followed by fusion of their long arms at the centromere. The short arms contain no essential genes, the individual is phenotypically normal and the genome contains the correct amount of genetic material (balanced translocation). One in 4 offspring will have the same balanced translocation, 1 in 4 normal chromosomes and 1 in 2 an unbalanced translocation.

Balanced translocation carriers occur in 1 in 250 couples contemplating pregnancy, 1 in 30 couples with a history of recurrent miscarriage and 1 in 5 of couples with a history of mixed fetal wastage. The risk of fetal abnormality depends on: (a) the sex of the carrier (15% risk of translocation trisomy 21 when the mother is the carrier, 3% with the father, because affected sperm have reduced functional ability); (b) the type of aberration (translocation or inversion); (c) the chromosome involved; and (d) the extent of chromosome imbalance.

Abnormalities of Chromosome Structure

These may be: (a) inversions, either pericentric (breaks on both sides of the centromere, commonly affecting chromosome 9) or paracentric (breaks at the end of one arm, resulting in duplication or loss of genetic material and an affected fetus); (b) de novo appearance of marker chromosomes (supernumerary chromosomal material) associated with a 1 in 4 risk of moderate to severe mental handicap; (c) ring chromosomes, with deletion of genetic material; (d) isochromosome (usually loss of the short arm of chromosome X); and (e) cell culture artefacts.

Single Gene Defects

The position of a gene on a chromosome is known as the locus; alleles are the different forms of that gene. An individual is homozygous with respect to an allele when both of a pair of chromosomes have identical alleles.

Mendelian inheritance (Table 7.2) may be either dominant or recessive. With dominant inheritance the gene is expressed in all persons possessing it and is transmitted to half the offspring of an affected parent; new mutations account for 50% (heterozygous form of achondroplasia) to 96% (tuberous sclerosis) of cases and are commoner in older fathers. Males affected by sex-linked dominant conditions (e.g. congenital hypophosphataemic rickets) will transmit the disease to all of their daughters. With recessive inheritance the condition is expressed only if the patient is homozygous. Once an affected child is born, there is a 1 in 4 recurrence risk rising to 1 in 2 if an affected patient marries a heterozygote. The risk of recessive conditions is higher in consanguineous marriages. Carriers of a sex-linked recessive condition (e.g. haemophilia, fragile X syndrome) will transmit this to 50% of their male offspring and 50% of their female offspring will be carriers. One third of haemophiliacs arise by a new mutation; females may be affected if a carrier marries an affected individual or if a carrier egg is fertilised by a mutant X sperm.

Table 7.2. Some clinically important single gene disorders: direct prenatal diagnosis using gene probes can be attempted for most

Inheritance	Disorder	Incidence (per 1000 live births)	Recurrence risk
Autosomal dominant	Monogenic hypercholesterolaemia	2	50% of offspring of affected parent
	Huntington's chorea	0.5	
	Neurofibromatosis	0.4	
Autosomal recessive	Cystic fibrosis	0.4	25% recurrence risk; 50% if affected parent marries heterozygote
	Sickle cell disease	0.1	
	Congenital adrenal hyperplasia	0.1	
	Inborn errors of metabolism e.g. phenylketonuria	0.01–0.02	
	Thalassaemia (alpha and beta)	Varies with ethnic group	
X-linked recessive	Red/green colour blindness	8	50% of male offspring of carrier
	Fragile X syndrome	1.0 (male)	
	Haemophilia (factor VIII deficiency)	0.1	
	Haemophilia (factor IX deficiency)	0.01	
	Duchenne's muscular dystrophy	0.2	
	Tay–Sachs disease	1 in 30 Ashkenazi Jews to a carrier	

Diagnosis of Genetic Defects

Preconception identification of carriers requires screening of certain ethnic groups: 1 in 30 Ashkenazi Jews is a carrier for Tay–Sachs disease; in London, 14% of the Cypriot population carries the beta thalassaemia gene; 9% of black Caribbeans and 17% of those of direct African origin carry the sickle cell gene.

Management includes: (a) genetic counselling in couples with a positive family history; (b) prenatal diagnosis (Table 7.3) using tissue obtained by placental biopsy (transcervically or transabdominally), amniocentesis, fetal blood or tissue sampling; and (c) abortion of affected fetuses or selective fetal therapy.

Problems may arise because of the heterogeneous amniotic fluid cell population derived from the fetus, extraembryonic membranes and trophoblast. Two cell lines with different karyotypes in the same amniotic fluid (2–17 cases per 1000 amniocenteses) may reflect true mosaicism in the fetus, with a risk of physical or mental handicap (20% of autosomal and 60% of sex chromosome mosaicism), pseudomosaicism from maternal cell contamination, or confined placental mosaicism. Repeat sampling of amniotic fluid is not generally helpful; it is best to analyse tissue derived solely from the fetus, e.g. fetal blood lymphocytes.

Environmentally Induced Abnormalities

Major causes are listed in Table 7.4. Of these only syphilis and rubella are routinely screened for antenatally.

Apart from the dose of the toxic agent, the outcome depends on the developmental state of the fetus at the time of exposure. Implantation, blastocyst formation and gastrulation occur in the first 17 days. A teratogen may cause miscarriage, resorption or survival intact by multiplication of the still totipotent cells. Organogenesis occurs from 18 to 55 days after conception: a teratogen may cause gross structural abnormalities because tissues are rapidly differentiating and damage is irreparable. From 56 days to term, defects of growth and functional loss may occur. There are species-related differences in the metabolism and disposition of toxic agents that limit the extrapolation of animal data to humans. The physiological changes of pregnancy (increased total body water, decreased binding proteins, increased liver metabolism and renal plasma flow) alter the distribution and elimination of drugs, resulting in lower plasma drug concentrations.

Multisystem Abnormalities

These are anatomical abnormalities due to the combined action of environmental and genetic factors, most of which are unknown. Over 90% occur in pregnancies without detectable risk factors. There is a 5% risk among first degree relatives and 25% concurrence in monozygotic twins. Risk increases with each affected pregnancy (Table 7.5).

Most are diagnosable with ultrasound (Tables 7.6–7.9), which is non-invasive and allows accurate pregnancy dating and early diagnosis of multiple pregnancy. However, accuracy is very dependent on the quality of the operator and the equipment.

Table 7.3. Invasive procedures for prenatal diagnosis

Tissue	Major indications	Advantages	Disadvantages
Amniotic fluid cells	Appreciable risk of serious genetic disease: 1. Raised maternal age 2. Previous abnormality 3. Abnormal biochemistry 4. Abnormal ultrasound findings (increased nuchal translucency)	1. Low risk: 1% procedure-related losses 2. High success rate 3. Low level of expertise required	1. Not earlier than 12–16 weeks 2. Culture takes 8–14 days 3. 0.06% karyotype discordance rate (maternal contamination and mosaicism) 4. Late termination of affected fetus 5. Increased risk of pulmonary hypoplasia inversely proportional to gestational age
Placenta	As above and diagnosis of: 1. Cytogenetic defects 2. DNA abnormalities (e.g. haemoglobinapathies) 3. Enzyme defects (e.g. Gaucher's and Tay–Sachs)	1. First trimester diagnosis 2. Direct DNA analysis and cytogenetic preparation (incl. sexing) in <24 h 3. Success rate >98% within three attempts 4. Small tissue volume required (15 mg chorionic villi) 5. Early termination possible	1. High level of expertise required 2. Long-term culture takes 7–10 days 3. Risks of sac perforation, bleeding, infection, isoimmunisation and limb/facial abnormalities (before 10 weeks) 4. 2% karyotype discordance rate (maternal contamination and mosaicism) but false negatives very rare with long-term culture 5. 1–2% procedure-related losses 6. High proportion of chromosomal abnormalities destined to miscarry (4% at 8 weeks versus 1% at 16 weeks) 7. Cost for cytogenetic indications double that of amniocentesis (both direct preparation and culture required) 8. Follow-up amniocentesis required in up to 10% of cases to ascertain diagnosis
Fetal blood	Rapid karyotyping in: 1. Severe fetal abnormality 2. Extreme growth retardation 3. Culture problems e.g. mosaicism, failed culture of conditions requiring special culture techniques (e.g. fragile X syndrome) 4. Diagnosis of blood disorders (haemoglobinapathies, immunodeficiencies and metabolic diseases)	1. Fetal karyotype can be obtained 2. Rapid result (24–48 h) 3. Acid-base status can also be obtained 4. Unnecessary obstetric interventions may be avoided	1. Needs skilled operator 2. Risks higher in non-immune hydrops 3. 2–3% risk of cord haematoma, haemorrhage, premature delivery, bradycardia, fetal death

NB: 8% of all pregnancies are sufficiently at risk to warrant prenatal diagnosis. Chorionic villus sampling is of greatest value in women at high risk of single gene defects or unbalanced chromosomal translocations.

Table 7.4. Environmental-induced abnormalities

Agent/dose	Effect
Alcohol	
1. 80 g/day before and during pregnancy	Fetal alcohol syndrome: (1) CNS abnormalities (mental retardation and microcephaly in 80%); (2) growth deficit with failure to catch up after birth; (3) abnormal facies, flattened profile, hypoplastic upper lip, etc; (4) congenital abnormalities in 25–50% (skeletal, cardiac, genital)
2. Moderate: up to 40 g/day	No clear evidence of congenital abnormality; contradictory data concerning spontaneous abortion and growth retardation
Tobacco	1. Dose-related increase in abortion, placenta praevia, abruption, premature rupture of membranes and perinatal death 2. No evidence of increase in congenital anomaly rate 3. Threefold increase in small-for-gestational-age babies with long-term effects on growth and intellectual development
Addictive drugs Heroin, methadone, cocaine (incl. crack)	1. No increase in incidence of severe congenital abnormalities (except cocaine) 2. 20–50% incidence of low birthweight babies 3. Fetal and neonatal dependence 4. Increase in sexually transmitted diseases (see below)
Therapeutic drugs	
Most anticonvulsants	Growth retardation; craniofacial anomalies (cleft lip/palate), fourfold increase in congenital heart disease, limb defects, mental retardation
Lithium	Growth deficiency; fivefold increase in congenital heart disease
Prochlorperazine	Teratogenic between 6–10 weeks
Benzodiazepines	No evidence of teratogenic effect
Warfarin	Chondrodysplasia punctata, CNS and face anomalies in 30% of patients treated at 6–11 weeks; brain and eye anomalies also in second/third trimesters
Beta blockers	No evidence of teratogenic effect; growth is at most slightly retarded
Vitamin A derivatives (isoretinion, etretinate)	CNS, craniofacial, cardiac abnormalities
Streptomycin	Deafness
Tetracycline	Discoloration and dysplasia of teeth and bones
Metronidazole	Theoretical risk of teratogenesis
Antiviral agents	Embryotoxic in animals, theoretically teratogenic in women
Aspirin	Inconclusive evidence for teratogenesis, affects platelet function and haemostatis in late pregnancy
Metoclopramide	Insufficient data in early pregnancy
Cytotoxic agents	Teratogenic and mutagenic
Radioisotopes	Thyroid damage with ^{131}I
Oral hypoglycaemics	Asplasia cutis with methimazole, fetal hypothyroidism
Sex hormones	
Non-steroidal oestrogens, e.g. stilboestrol	Adenocarcinoma of vagina and cervix
Steroidal oestrogens	Inconclusive teratogenicity data
Medroxyprogesterone and danazol	Virilise female fetus
Cyproterone acetate	Inhibits masculinisation of male fetus in animal studies only
Bromocriptine	No adverse effects
Bacterial infection	
Syphilis	Abortion, fetal death, growth retardation, congenital anomalies, defective bone growth, anaemia, purpura, rashes, CNS damage
Listeria monocytogenes	Infects 1 in 20 000 births; abortion, stillbirth, premature labour, congenital infection presents with pneumonia, septicaemia after birth (granulomatosis infantiseptica)

Table 7.4. (continued)

Agent/dose	Effect
Viral infection (clinical illness in 5% of pregnancies)	
Rubella	First trimester: 10–50% congenital abnormalities (cardiac, cataract, deafness, microcephaly), mental retardation, death
	Second and third trimesters: 4% risk of deafness, mental retardation
Cytomegalovirus (primary infection or reactivation of latent virus) – 0.3% of live births	2% serious: CNS abnormalities (calcification, microcephaly), congenital heart disease
	8% minor/transient: hepatosplenomegaly, jaundice, petachiae, purpura
	90% symptom free at birth, later may develop deafness, low IQ
Herpes virus	CNS abnormalities (atrophy, microcephaly, chorioretinitis); 30% of infected neonates have neurological disease
Varicella zoster	Questionable teratogen; congenital chicken pox (fever, rash, respiratory disease) is often lethal; increased risk of malignancy in childhood
Coxsackie	Questionable teratogen
Influenza, polio, measles, mumps	Increased risk of abortion; questionable teratogen
Hepatitis B	First/second trimester: 10% fetal infection (liver damage)
	Third trimester: 75% fetal infection (liver damage)
	Increased risk of prematurity and growth retardation
Human immunodeficiency virus	Neonatal AIDS
Toxoplasma gondii	1 in 10 000 women in UK; 20% fetal infection in first trimester; clinically indistinguishable from CMV except for periventricular calcification
Other causes	
Mercury	Congenital encephalopathy
Lead	Congenital encephalopathy, behavioural effects, decreased hearing
Radiation	Dose-dependent increased incidence of malignancy and microcephaly; germ cells, blood cells and epithelia. Radiation damage can be repaired

CNS, central nervous system; CMV, cytomegalovirus.

Table 7.5. Clinically important multisystem abnormalities

Type	Incidence	Features	Recurrence	Diagnosis
1. NTD (50% of all congenital anomalies)				
a. 50% anencephaly	3–9/1000 live births in UK	50% survive to age 5	After 1 child, 1 in 20; after 2 children, 1 in ten; after 3, 1 in 5	At 16–18 weeks, AFP > 2.5 MoM: 90% anencephaly, 80% open spina bifida, 3% normals (false positives)
b. 45% spina bifida	Wide temporal, ethnic, geographic variation Females > Males	85% of these severely handicapped (paralysis, urinary/faecal incontinence, mental retardation)	First degree relatives have 10 × background risk	False negatives: < 10% for anencephaly, < 20% for open spina bifida
c. 5% encephalocoele			Preconception multivitamins reduce recurrence	False positives: laboratory error, wrong dates, multiple pregnancy, omphalocoele, gastroschisis, congenital nephrosis, obstructive uropathy, cystic hygroma, Turner's syndrome, trisomy 13 Ultrasound scan: false positive < 1%: false negative < 1% for anencephaly/encephalocoele; false negative < 20% for spina bifida
2. CHD (50% are major anomalies e.g. hypoplastic left heart, pulmonary stenosis, aortic coarctation, transposition, Fallot's)				
	4–8/1000	25% are fatal	One affected child, 1 in 50	US: four-chamber view excludes major CHD in 90% of cases by 20 weeks
	Higher if other structural anomalies present	Fetal heart failure results in hydrops (ascites, pleural effusion, skin oedema)	One affected parent, 1 in 10	Remainder require detailed US, fetal echo and colour flow mapping
		Maternal digitalisation is effective in fetal dysrythmia		

Congenital Abnormalities

Table 7.5. (continued)

Type	Incidence	Features	Recurrence	Diagnosis
3. GIT abnormalities				
a. Oesophageal	2–10/10 000	90% have tracheo-oesophageal fistula 50–70% have other abnormalities (mainly cardiac) 3–4% have chromosomal abnormalities 10% have obstruction complete (therefore detectable)		For US diagnosis, see Table 7.7
b. Duodenal atresia	1 in 10 000	30% have trisomy 21 associated malformations		For US diagnosis, see Table 7.7
c. Diaphragmatic hernia	2–5/10 000	25% have other abnormalities 25% have chromosomal abnormalities		For US diagnosis, see Table 7.7

NTD, neural tube defects; CHD, congenital heart disease; GIT, gastrointestinal tract; US ultrasound; AFP, alphafetoprotein; MoM, multiples of the median.

Table 7.6. Clinical and ultrasound features of major CNS abnormalities

Anomaly	Clinical features	Major US features
Neural tube defects		
1. Anencephaly	Failure of neural tube closure; incompatible with postnatal life; may cause polyhydramnios	Absent cranial vault; readily detected even in first trimester
2. Meningocoele	Subtype of spina bifida	Defect in skin cover ± abnormal neural arch ossification centres; spinal cord intact, BPD and HC often small for gestational age; lemon and banana signs
3. Meningomyelocoele	Prognosis worse when associated with hydrocephalus	Bony defect of spine and disruption of cord; normal limb movements and bladder filling
4. Encephalocoele	Defect of cranial vault, usually occipital; if brain herniates (exencephaly), prognosis is poor; usually isolated (recurrence 1 in 20); may be part of Meckel's syndrome (with multicystic dysplastic kidneys, autosomal recessive inheritance)	Bony defect may be small; soft tissue mass may be large
Microcephaly	Inadequate brain growth, subnormal IQ; multifactorial: infections, radiation, mercury, phenytoin; 3% recurrence except in families with autosomal recessive inheritance	Fall off in all skull diameters relative to other measures, usually late in second trimester; accurate dating crucial
Hydrocephaly	80% of open spina bifida have hydrocephaly; 3% recurrence if isolated; aqueduct stenosis (sex-linked linked recessive, 1 in 2 recurrence in males); poor prognosis; 85% have associated abnormalities, 20% chromosomal abnormalities	BPD not increased in early stages; ventricle: hemisphere ratio > 0.5 after 18 weeks (5% false positives); definitive diagnosis requires serial measurements; limited role for in utero shunting
Brain cysts	Choroid plexus cysts disappear by 25 weeks in 90% of cases; porencephalic cysts communicate with cerebral ventricles; prognosis variable	Cystic structures in brain substance are easily distinguishable
Iniencephaly	Fusion of occiput and cervical vertebrae usually in females	Gross neck hyperextension; cross-section of head and thorax in same plane
Holoprosencephaly	Defect in formation of midline structures; 20% have associated chromosomal defects	A single dilated midline cerebral ventricle and associated facial defects
Hydroencephaly	Congenital absence of cerebral tissue: carotid artery atresia, necrotising encephalitis, trisomy 13/15	Complete absence of echoes from cerebral vault
Posterior fossa abnormalities	Absent cerebellar hemispheres (in trisomy 5/18); absent vermis and enlarged cisterna magna (Jouberts syndrome); Dandy–Walker malformation; 50% have chromosomal defects in presence of additional malformations	US diagnosis possible using suboccipito-bregmatic view

NB: Additional abnormalities increase the probability that the fetus is chromosomally abnormal. Karyotyping is then indicated.
BPD, biparietal diameter; HC, head circumference; US, ultrasound.

Congenital Abnormalities

Table 7.7. Clinical and ultrasound features of major gastrointestinal and abdominal wall anomalies: frequently associated with other major structural/chromosomal anomalies

Abnormality	Clinical features	Major US features
Gastrointestinal anomalies		
1. Oesophageal atresia	Polyhydramnios	Polyhydramnios and stomach bubble absent; demonstrable only when obstruction is complete (10% of cases)
2. Duodenal atresia	30% have trisomy 21; 50% associated anomalies	Double bubble
3. Jejunal obstruction		Many dilated loops with peristalsis
4. Anal agenesis and Hirschprung's disease		Multiple irregular distended bowel loops, esp. left flank
5. Diaphragmatic hernia	Through foramen of Bochdalek, usually left sided; easily repaired; if large, neonatal death due to pulmonary hypoplasia	Stomach bubble (90%) or liver (50%) adjacent to heart in thorax; mediastinal shift; polyhydramnios
Abdominal wall defects		
1. Omphalocoele (Exomphalos)	2–4/10 000 births; failure of gut to return to abdominal cavity after 11 weeks; 50% have cardiac/chromosomal anomalies; elevated AFP; 30% neonatal mortality after surgical correction, recurrence rate < 1%	Midline abdominal wall defect, size variable; sac may contain small bowel, stomach, colon, liver with peritoneal cover
2. Gastroschisis	1/10 000 births; extruded bowel without peritoneal covering; 10–30% associated with anomalies; rarely chromosomal; 15% neonatal mortality after surgical correction	Small defect in anterior abdominal wall, usually to the right of umbilicus, free floating small bowel

US, ultrasound; AFP, alphafetoprotein.

Table 7.8. Clinical and ultrasound features of major renal tract abnormalities. Renal tract abnormalities occur in 2–3/1000 pregnancies; chromosomal defects are found in 5% (renal malformations only) to 40% (with additional anomalies) of cases

Abnormality	Clinical features/prognosis	US features
Renal agenesis	Associated with anhydramnios, pulmonary hypoplasia, compression deformity; one cause of Potter syndrome (when bilateral)	Absent renal echoes; severe symmetrical growth retardation; two vessels in cord; diagnosis hampered by anhydramnios and fetal flexion
Congenital renal cystic disease		
1. Infantile form (Potter type I)	Autosomal recessive; fatal in childhood	Oligohydramnios; increased renal size and echogenicity (relative to lungs) due to microscopic cysts; may not become apparent before 26 weeks
2. Adult form (Potter type II)	Autosomal dominant; cysts present before birth but manifest in adults; may be part of Meckel's syndrome (with encephalocoele)	Renal cysts; may be indistinguishable from 1 and 3
3. Multicystic kidney	Prognosis generally good if liquor volume and other kidney normal	Unilateral renal cysts usual

Table 7.8. (continued)

Abnormality	Clinical features/prognosis	US features
Obstructive uropathy		
1. Urethral agenesis, stenosis or atresia	Low complete obstruction in males, usually fatal	Large bladder, hydroureter, hydronephrosis; severe oligohydramnios and dysplastic kidneys
2. Posterior urethral valve	Intermittent or incomplete obstruction in males; poor prognosis due to renal failure and pulmonary hypoplasia; 50% have associated chromosomal/structural anomalies	Large bladder; oligohydramnios less severe if obstruction intermittent; in utero bladder drainage is possible (vesicoamniotic shunting)
3. Pelvi-ureteric junction obstruction	Excellent prognosis, pyeloplasty may be necessary	Hydronephrosis in the absence of dilated ureters and bladder; normal renal cortical thickness and liquor volume

Table 7.9. Clinical and ultrasound features of major limb deformities. Most skeletal dysplasias have high perinatal mortality due to small chest size and pulmonary hypoplasia. They may be symmetrical or asymmetrical. Associated abnormalities and chromosomal defects are common. Limb reduction defects are associated with CVS at < 10 weeks' gestation

Abnormality	Clinical features	US features
Phocomelia	Small limb buds only; associated with diabetes, drugs, radiation	Limbs absent
Achondroplasia	Non-lethal heterozygote or lethal homozygote	Limbs short late in second trimester; early severe limb and chest reduction
Osteogenesis imperfecta	Congenital lethal form is usually autosomal dominant; tarda: only detectable in late pregnancy	Severe symmetrical limb shortening and deformity; poor calcification; intrauterine fractures
Hypophosphatasia	Inadequate calcification; lethal due to weakness of chest wall	Fetal bones transonic
Other forms of dwarfism	Most have autosomal inheritance, but thanatophoric dwarfism has 1 in 100 recurrence risk	Prominent limb reduction deformity

CVS, chorion villus sampling; US, ultrasound.

Screening for Congenital Abnormalities

The most widely used screening programmes are listed in Table 7.10.

Table 7.10. Common screening programmes in pregnancy

Who?	What?	When?	Result?	What next?
All pregnant women after informed consent	Serum AFP, uE3, hCG and US (BPD)	15–18 weeks	1. AFP > 2.5 MOM: 80% detection rate and 3% false positives for open spina bifida 2. US: 100% detection rate for anencephaly; no false positives 3. Triple test and maternal age: risk of Down's > 1 in 250 (61% detection rate; 5% false positives for Down's)	Amniotic fluid AFP and anomaly scan (97% detection rate and 0.4% false positive rate for open spina bifida) TOP or await delivery depending on gestational age Amniocentesis; offer TOP after counselling
	Serum hCG (beta subunit); PAPP-A	10–13 weeks	As for triple test	CVS; offer TOP after counselling
	Nuchal translucency	10–13 weeks	> 3 mm (80% detection of Down's syndrome; 5% false positives)	CVS; offer TOP after counselling
Woman with cystic fibrosis or known carrier; previous affected child	CF gene probe	First trimester	If positive	Offer TOP after counselling
All black women; previous child with sickle cell disease	Haemoglobin electrophoresis in both partners at booking	First trimester	If both HbS carriers	Genetic counselling and CVS (T1) or fetal blood sampling (T2/3) for beta globin gene probe analysis; offer TOP if positive
Mediterranean and Asian origin; previous child with thalassaemia	Haemoglobin electropheresis in both partners at booking	First trimester	If both thalassaemia trait	Genetic counselling and CVS (T1) or fetal blood sampling (T2/3) for thalassaemia gene probe analysis; offer TOP if positive

TOP, termination of pregnancy; CVS, chorion villus sampling; AFP, alphaprotein; hCG, human chorionic gonadotrophin; US, ultrasound; BPD, biparietal diameter; CF, cystic fibrosis; PAPP-A, pregnancy-associated plasma protein-A; uE3, unconjugated oestriol; MOM, multiples of the median. T1, first trimester; T2, second trimester; T3, third trimester.

Chapter 8

Rhesus Disease

The sensitisation of a rhesus-negative mother to rhesus-positive blood occurs in 10% of deliveries (affecting 0.5% of their infants), 1–2% of first pregnancies (secondary to previous transfusion, amniocentesis, chorionic villus sampling, external cephalic version or bleeding), 3–4% of spontaneous miscarriages, and 5–6% of induced abortions. Maternal immunoglobulin (Ig) G antibodies cross the placenta, destroying fetal red cells by haemolysis, producing fetal anemia and stimulating bone marrow production at extramedullary sites, primarily liver and spleen. The term erythroblastosis is derived from the presence of nucleated red cell elements in the fetal circulation.

There are five main types of rhesus antigen, called D, C, E, e and c; the D antigen, followed by c and E, causes the most severe haemolytic reactions. Approximately 8% of the black population lack the D antigen (rhesus negative), about half the incidence in whites. Only rhesus-positive fetuses will cause rhesus immunisation, and will be affected by the rhesus antibody produced. ABO incompatibility between mother and fetus reduces the risk of sensitisation by one tenth, because of the rapid intravascular haemolysis of fetal cells that are coated with anti-ABO antibodies. Apart from the rhesus and ABO systems, there are many other antigens which most commonly give rise to anti-Lewis antibodies, though they are not haemolytic.

Diagnosis

Identification of severely affected fetuses is based on (a) obstetric history (50% risk of intrauterine death if the previous fetus was severely affected; 70% risk after previous intrauterine death); (b) the levels and trend of maternal rhesus antibody (4–20 IU/ml) indicates mild haemolysis; more than 20 IU/ml indicates severe haemolysis; serial measurements predict severity with 95% accuracy); (c) after 24 weeks, amniotic fluid bilirubin levels (using spectrophotometry at a wavelength of 450 nm and Liley's chart); (d) percutaneous umbilical fetal blood sampling to determine whether fetal haemoglobin and haematocrit are less than two SD below the mean for gestation; and (e) detection of hydrops and its severity by serial ultrasound scans.

Management

Packed maternally compatible blood is transfused via an umbilical vessel (under ultrasound control) in proportion to the fetoplacental blood volume and the pretransfusion and transfused blood haematocrit. There is a 1% drop in fetal haemoglobin per day. Transfusion is repeated every 1–3 weeks to maintain the fetal haematocrit above one third of the normal mean for gestation. Treatment can be continued until delivery. Antenatal rhesus prophylaxis at 28 weeks is recommended in the USA.

After delivery, a Kleihauer test (to estimate the amount of fetomaternal haemorrhage and determine the dose of rhesus immune globulin to be given), fetal haemoglobin, haematocrit, blood group and Coombs' test are performed on cord blood. If the infant is rhesus positive, sensitisation is prevented by administration of 100 μg anti-D Ig as soon after delivery as possible. This is sufficient to neutralise 4 ml of fetal blood, equivalent to 80 erythrocytes per 50 low power fields (Kleihauer test). Additional doses should be given to clear all fetal erythrocytes from maternal blood. Prophylaxis should be given, even if delayed, though it may not be effective.

Prognosis

This depends on severity. One half of affected babies do not require treatment (mild). In moderate cases (25–30% of affected infants), slight anaemia and jaundice are treated by phototherapy or exchange transfusion with maternally compatible rhesus-negative blood. Untreated severe disease (fetal haemoglobin less than one third of mean for gestation) causes generalised fetal oedema, ascites, hydrothorax (hydrops), intrauterine death (15%) or kernicterus (90% fatal) in the infant. In about half the cases, the prognosis worsens with each subsequent pregnancy. Intravascular transfusion gives an 80% or greater survival rate.

Chapter 9

Hypertensive Disorders of Pregnancy

In normal pregnancy, blood pressure falls to a nadir in the second trimester (average systolic and diastolic blood pressure 5 and 15 mmHg (1 mmHg = 133.3 Pa) lower than prepregnancy values, respectively), rising to prepregnancy levels in the third trimester. Blood pressure should be measured in the sitting or lateral position with the sphygmomanometer at the level of the heart, using a cuff wide enough to cover 80% of the arm circumference. Hypertension is defined arbitrarily as diastolic blood pressure of 90 mmHg or more (muffling of Korotkoff sounds, Point IV) throughout pregnancy. This corresponds to 3 SD above the mean in the first two trimesters and 2 SD above the mean in the third; above this level there is a significant increase in perinatal mortality. The true incidence of hypertension is difficult to establish: overall one third of all maternities are complicated by hypertension, approximately half of these occurring for the first time in labour.

Classification

Six clinical groups are described:

1. *Pregnancy-induced hypertension* in previously normotensive non-proteinuric women: it may be gestational hypertension (without proteinuria), pre-eclampsia (with proteinuria) or eclampsia (with fits).
2. *Chronic hypertension* in women known to have hypertension or chronic renal disease: it may be essential hypertension (without proteinuria), chronic renal disease (with proteinuria) or other causes (e.g. phaeochromocytoma, coarctation of the aorta).
3. *Pre-eclampsia or eclampsia* (with fits) superimposed on gestational or chronic hypertension.
4. *Transient intrapartum hypertension* occurring during labour or in the puerperium, returning to normal within 48 h of delivery.
5. *Unclassified*: late hypertension with or without proteinuria occurring in a patient booking after 20 weeks' gestation or without sufficient data to permit classification.

6. *Eclampsia*: generalised convulsions during pregnancy, labour or within seven days of delivery, not caused by epilepsy or other convulsive disorders.

Pregnancy-Induced Hypertension

Hypertension in normotensive pregnant women may be due to pre-eclampsia or gestational hypertension. Mild cases are defined by a spot diastolic blood pressure of 100–120 mmHg and a repeat diastolic pressure (after 6 h) of 90–110 mmHg in the absence of proteinuria or maternal complications. Renal function and fetal growth are normal. In severe cases, spot diastolic blood pressure is greater than 120 mmHg, spot systolic greater than 160 mmHg, repeat diastolic greater than 110 mmHg and repeat systolic greater than 160 mmHg. Proteinuria is greater than 300 mg/24 h and fetal growth or condition may be impaired.

Gestational Hypertension

This is hypertension without proteinuria in previously normotensive women. It affects multigravidae more than primigravidae; the frequency and severity increase with maternal age. It often recurs, is familial and is thought to be due to an inherited latent hypertensive tendency. These women have a higher incidence of hypertension in later life. Eclampsia is very rare without proteinuria. The typical glomerular lesions of pre-eclampsia are absent.

Pre-eclampsia

This is hypertension with proteinuria in previously normotensive women. It complicates 5% of all pregnancies (10% of primigravidae). It may recur and is familial. Maternal age but not social class or smoking are risk factors; short, obese women are at greater risk; fetal risk factors include multiple pregnancy, hydrops fetalis and hydatidiform disease. Pre-eclampsia is not associated with an increased incidence of hypertension in later life except in group 3 patients.

The aetiology is unknown: genetic factors together with an abnormal immunological reaction to the first pregnancy have been postulated. There is failure of the second wave of normal trophoblast invasion of the spiral arteries, and uteroplacental blood flow is impaired. Reduced trophoblast volume in the spiral arteries leads to imbalance of the prostacyclin–thromboxane system, which causes local platelet aggregation, vasospasm of the spiral arteries and removal of pregnancy protection against circulating pressor agents such as angiotensin II. The retained muscle coats are sensitive to angiotensin II and peripheral resistance does not fall.

Acute atherosis (fibrin and platelet deposition) in terminal segments of the uterine spiral arteries is characteristic. The subsequent rise in blood pressure further reduces placental perfusion, leading to fetal growth retardation, hypoxia and intrauterine death. Activation of the complement system causes immune complexes to be deposited on the basement membrane of the kidney. The key pathological feature is swelling of the glomerular endothelial cells with occlusion of capillary lumina.

Pre-eclampsia is a multisystem disorder. It affects:

1. *The maternal arterial system*: hypertension is characterised by unstable basal blood pressure with reversed circadian pattern and spikes unrelated to external stimuli. Sudden elevations of blood pressure can cause acute arterial damage and loss of vascular autoregulation. Cerebral haemorrhage is the commonest cause of maternal death from eclampsia and pre-eclampsia.

2. *The renal system*: proteinuria (24 h total protein excretion greater than 300 mg) occurs in 5–15% of women with hypertension in pregnancy and indicates poor prognosis for mother and baby. The decline in renal function is biphasic: tubular function impairment (serum urate greater than 0.35 mg/dl) precedes decreased glomerular filtration (plasma creatinine greater than 90 µmol/l and urea greater than 6 mmol/l). Though not specific to pre-eclampsia, decreased urate excretion is a sensitive early indicator of deteriorating renal function and identifies a subgroup of women with a 10-fold increase in perinatal mortality rate.

3. *Coagulation system*: elevated α2 macroglobulin and decreased antithrombin III activity activate the clotting system, increasing factor VIII consumption (compensated activation). Decompensation with widespread fibrin deposition is characteristic of disseminated intravascular coagulation; complications include renal cortical and periportal hepatic necrosis and microangiopathic necrosis (haemoglobinaemia, reduced haptoglobin and red blood cell fragmentation).

4. *Liver involvement*: increased hepatic enzymes, jaundice, periportal haemorrhage, necrosis and rarely hepatic rupture (epigastric pain and circulatory collapse not always preceded by jaundice, with a maternal mortality of 60%).

Screening for pre-eclampsia by means of the roll-over test (increase in diastolic blood pressure of greater than 20 mmHg on turning from the lateral to supine position) is not widely used because of high false-positive and false-negative rates and poor reproducibility. Doppler screening of the uteroplacental circulation in the second trimester identifies a group of women with a loss of end-diastolic flow. These women have a 10-fold increased risk of developing pregnancy complications (perinatal death, pre-eclampsia, growth retardation and abruptio placentae), but overlap between pregnancies with normal and abnormal outcome does not justify its routine use.

Management of mild cases includes decreasing physical activity, weekly blood pressure measurements and baseline investigation of renal function, haematology screen with platelet count, fetal growth and well-being assessment (kick chart, CTG). Bed rest is of no proven benefit in non-proteinuric hypertension. Hospitalisation should be limited to symptomatic women (headaches, etc.), asymptomatic but proteinuric women and asymptomatic, non-proteinuric women with elevated urate excretion; the remainder can be managed as outpatients.

Once hospitalised, severe pre-eclamptic women should have weekly renal and liver function tests as well as platelet count, coagulation status and protein excretion. Elevated urate levels and falling platelet counts reflect worsening of the clinical condition. Fetal size and liquor volume are assessed ultrasonically; CTG is repeated at intervals determined by clinical status; absent or reversed flow in the umbilical arteries during diastole as determined by Doppler ultrasonography is associated with higher perinatal morbidity and mortality.

The definitive treatment of severe pre-eclampsia is delivery of the fetus. Antihypertensive treatment (to protect the mother from the risk of cerebral haemorrhage, left ventricular failure, renal failure, disseminated intravascular coagulation and convulsions) is indicated when the mean arterial pressure exceeds the threshold for vessel injury, i.e. 140 mmHg. These drugs have no effect on the progression of the disease.

Therapeutic options include the following.

1. *Beta adrenergic blockade* (e.g. propranolol). This decreases cardiac output and hence blood pressure. Side-effects include decreased uterine and renal blood flow, and premature labour (only at high doses). Placental transfer of the drug may lead to hypoglycaemia, bradycardia and apnoea, especially in premature or growth-retarded neonates. Labetalol blocks both alpha and beta receptors, acting as a vasodilator and decreasing uterine activity; uteroplacental blood flow and FHR are not altered. The onset of the antihypertensive effect is rapid.
2. *Vasodilators* (e.g. hydrallazine, magnesium sulphate, diazoxide and clonidine). Autoregulation compensates for the fall in renal and uterine (radial and arcuate branches) blood flow unless the hypotension is precipitous. Maternal side-effects of hydrallazine include headaches, vomiting, shakiness, hyperreflexia (mimicking impending eclampsia), tachycardia and renin stimulation (abolished by beta blockers). Women who are slow acetylators have a genetically decreased capacity to metabolise the drug; a rare systemic lupus erythematosus type syndrome is described at doses in excess of 300 mg/day. No serious fetal side-effects have been reported. The effectiveness is increased by combination with methyldopa because sympathetic inhibition blocks the reflex tachycardia secondary to the blood pressure drop.
3. *Central alpha receptor stimulation* (e.g. methyldopa). This effectively lowers blood pressure by reducing sympathetic nervous system outflow; it has no effect on renal, haemostatic or placental function. Maternal side-effects include fatigue, depression, dizziness and nightmares; minor fetal side-effects include transient low heart rate and blood pressure after delivery.
4. *Calcium channel blockers* (e.g. nifedipine).
5. *Epidural analgesia* in labour counteracts hypertension by blocking the release of catecholamines, and improves intervillous and renal perfusion by relieving vasoconstriction of the spiral and renal arteries. Fluid preloading is essential to avoid a precipitous fall in blood pressure induced by the vasomotor block.

Some women at high risk for pre-eclampsia may benefit from low-dose aspirin (75 mg/day) in the prenatal period. Aspirin irreversibly acetylates the platelet cyclo-oxygenase enzyme and this inhibits thromboxane production. This prevents acute atherosis.

Eclampsia

Epigastric pain, headaches, visual disturbances, hyperreflexia, rapidly developing generalised oedema, deteriorating renal function with urine output less than 20 ml/h may indicate imminent eclamptic fits; it occurs in fewer than 1% of cases of correctly managed severe pre-eclampsia. Management includes: (a) control of the fits with magnesium sulphate or phenytoin ; (b) reduction of blood pressure with hydrallazine, diazoxide or sodium nitroprusside (exceptionally); (c) maintenance of fluid, electrolyte and acid–base balance by monitoring central venous pressure, urea and electrolytes and blood gases; (d) delivery; and (e) treatment of complications, e.g. abruption, cardiac and renal failure.

Management of Fluid Retention in Pregnancy

Oedema occurs in 80% of normal pregnant women; it has no diagnostic or prognostic significance. It occurs in 85% of women with pre-eclampsia, but "dry" pre-eclampsia has a worse prognosis. Ascites may occur in severe pre-eclampsia. Rapidly developing generalised oedema may indicate imminent eclampsia. Pulmonary oedema is rare; it may be due to cardiac failure secondary to hypertension, to fluid overload or to adult respiratory distress syndrome following prolonged hypoxia. Oedema can occur at relatively lower venous pressure levels in pregnancy due to the characteristically reduced oncotic pressure; treatment is with intravenous frusemide with central venous pressure monitoring.

Salt restriction is contraindicated in the treatment of fluid retention in pregnancy because it aggravates renal impairment and is associated with increased perinatal mortality. Plasma volume expansion may cause circulatory overload and pulmonary oedema. Calorie restriction, though not altering the incidence of pre-eclampsia, causes a significant fall in birthweight.

Prophylactic diuretics do not prevent pre-eclampsia. Since hypovolaemia is characteristic of severe pre-eclampsia, any further contraction in plasma volume will exacerbate renal and placental blood flow impairment. If oliguria supervenes, and the central venous pressure is greater than 10 cm H_2O (1 cm H_2O = 0.1 Pa), 40 mg of frusemide will induce a diuresis if dehydration is the underlying cause, as will 200 ml of 5% dextrose if the central venous pressure is between 5 and 10 cm H_2O. In renal failure, diuresis does not occur; this condition is treated by fluid and protein restriction, hydrallazine in hypertensive women (because it increases renal blood flow) and early recourse to haemodialysis.

Prognosis of Hypertensive Disease in Pregnancy

The two major causes of maternal mortality are hypertension and pulmonary oedema. True pre-eclampsia (i.e. with proteinuria) is associated with a perinatal mortality double that of gestational hypertension, especially if patients present early. This is due to the consequences of growth retardation and the effects of iatrogenic preterm delivery.

The incidence of pre-eclampsia is increased two- to sevenfold in women with chronic hypertension: these pregnancies are associated with a higher perinatal mortality rate than pre-eclampsia in previously normotensive women.

There is a five- to tenfold increase in the perinatal death rate in eclampsia, depending on whether it occurs antenatally or in labour, and on the care given. The incidence of eclampsia is falling as are the number of maternal and fetal deaths due to this complication.

Chapter 10

Antepartum Haemorrhage

Antepartum haemorrhage (APH) occurs in 3% of pregnancies. Some 7% of maternal deaths and 17% of perinatal deaths are associated with APH. The definitions and classification are shown in Table 10.1. The practical clinical classification is placenta praevia, abruptio placentae and "indeterminate".

Abruptio Placentae

Factors that have been associated with severe abruptio placentae include high parity, hypertension, smoking, cocaine use, premature rupture of membranes, pre-eclampsia, trauma (including external cephalic version), uterine abnormalities, short umbilical cord, dietary deficiencies and sudden uterine decompression. Of these factors, hypertension is the commonest. The frequency of severe abruption is decreasing. Perinatal mortality is 25–35%. There is a high risk of recurrence in subsequent pregnancies (1 in 10–20).

Table 10.1. Classification of antepartum haemorrhage (APH); though traditionally defined as bleeding after the 14th week of pregnancy, the same conditions frequently occur in mid-trimester

Term	Definition
True	Bleeding from placental site
False	Bleeding from lower genital tract
Inevitable (placenta praevia)	Bleeding from low-lying placenta
Accidental	Bleeding from normally sited placenta
Abruptio placentae (placental abruption)	As accidental
Concealed accidental	Blood is retained between the detached placenta and uterus
Revealed accidental (marginal)	Bleeding from edge of normally sited placenta via cervix and vagina
Mixed accidental	Combination of retained blood and vaginal bleeding
Vasa praevia	Bleeding from fetal vessels
Indeterminate	Retrospectively defined as marginal, false, inevitable or accidental

The basic pathology is a split in the decidua basalis leading to rupture and haemorrhage from the decidual spiral arteries.

The clinical features are increasing and continuous abdominal pain and sometimes vaginal bleeding. The uterus is hard and tender; fetal parts may be difficult to palpate; the fetus may be dead or in severe distress; and there may be signs of hypotensive shock (contrary to earlier thinking, this is proportional to blood loss). Sometimes the patient is in labour; the contractions may be hypertonic with a high baseline pressure (25–50 mmHg). Atypical or even silent presentations are not uncommon. Complications of the most severe cases include renal failure (tubular or cortical necrosis), disseminated intravascular coagulation, fetal-maternal haemorrhage and extravasation of blood throughout the myometrium and under the serosa (Couvelaire uterus). Abruption with concealed haemorrhage carries a much greater risk for the mother. There is also an increased incidence of postpartum haemorrhage.

The haematoma can sometimes, but not always be identified by ultrasound. The main purpose of sonography is to exclude placenta praevia.

Minor cases of abruption can be managed conservatively, with close observation and induction at 38 weeks. More severe cases should receive immediate transfusion with 1 litre or more of blood, the patient's response being monitored by the central venous pressure (10 cm H_2O in the third trimester), haematocrit (30% or above) and urine flow (30 ml or more). If the fetus is alive it should be delivered by caesarean section, provided maternal clotting is effective. If it is dead then early delivery is achieved with amniotomy and oxytocin infusion (unless haemorrhage is uncontrollable).

Placenta Praevia

The incidence of placenta praevia is 1 in 150–250 births. It can be divided into four grades:

1. *Lateral*: placenta extends onto lower segment.
2. *Marginal*: placenta extends to the edge of the internal os.
3. *Complete acentric* (partial): placenta covers internal os but is not centrally placed.
4. *Complete centric* (total): placenta covers internal os and centrally placed.

Placenta praevia is associated with high parity, advanced maternal age, previous caesarean section, the large placenta seen in multiple pregnancy and erythroblastosis. The recurrence rate is 4–8%.

The commonest presentation is painless bleeding at 30 weeks. Subsequent bleeds are usually more severe than the first. The presenting part is high and abnormal lies (breech, transverse, oblique) are common. Bleeding may also occur in mid-trimester; this is associated with a very high fetal mortality (50% or more). The fetus is at risk of prematurity, growth retardation, and cerebral palsy. The position of the placenta is ascertained by ultrasound. This has an accuracy of 95% or better, especially if the placenta is anterior. Vaginal ultrasound may be superior to abdominal for posterior placentas. If a low-lying placenta is identified in early pregnancy, the scan should be repeated at 32 weeks. Most will "migrate" from the lower segment but central low-lying placentas remain low.

Vaginal (including speculum) and rectal examinations are not performed until placenta praevia is excluded.

Patients who are in labour with little bleeding and the presenting part in the pelvis can be delivered vaginally. All others should be delivered by caesarean section. After 37 weeks this is performed without delay. Before 37 weeks the patient should remain in hospital until the fetus matures. If severe bleeding continues the patient must be delivered regardless of fetal maturity. Abnormal adherence of the placenta to the lower segment (and mobility of the lower segment to contract adequately) may result in postpartum haemorrhage.

"Indeterminate" APH

"Indeterminate" bleeding, if it is not from the lower genital tract, may come from the edge of a circumvallate or "extrachorial" placenta (marginal APH). Provided the bleeding stops and there is no ultrasound evidence of placenta praevia or abruptio placentae, the woman can be managed as an outpatient. Bleeding from fetal vessels is diagnosed by testing the blood for fetal haemoglobin.

Disseminated Intravascular Coagulation (DIC)

This may be associated with abruptio, amniotic fluid embolism, missed miscarriage and long-standing hydatidiform mole and endotoxic shock. It is seen in 30% of women with abruption severe enough to kill the fetus. The deposition of fibrin in small vessels in a variety of organs leads to specific organ damage (especially the kidneys) and general depletion of coagulation factors (hypofibrinogenaemia and thrombocytopenia). The coagulation process leads to activation of plasminogen to plasmin; plasmin lyses fibrinogen and fibrin, with the formation of a series of "degradation" or "split" products; these products delay fibrin polymerisation and cause a defective fibrin clot structure. The condition is diagnosed by the whole blood clotting time (normal 5–10 min) and the thrombin clotting time (normal 10 s). Cases of DIC should be managed in consultation with a haematologist. Apart from emptying the uterus, the treatment (severe cases only) is liberal transfusion with fresh frozen plasma and platelets. Specific therapy with agents such as fibrinogen, heparin and ϵ-aminocaproic acid (EACA) is no longer recommended.

Chapter 11
Disorders Involving Amniotic Fluid

Hydramnios

The volume of amniotic fluid increases progressively (10 weeks, 30 ml; 20 weeks, 250 ml; 30 weeks, 750 ml; 36 weeks, 900 ml) and then falls slightly at term. Hydramnios is an excess of amniotic fluid (more than 2000 ml), usually accumulating gradually. Specific causes include: (a) unknown (in two thirds of patients); (b) fetal abnormalities, neural tube defects, oesophageal and duodenal atresia, congenital heart defects, Down's syndrome, alpha thalassaemia; (c) diabetes, usually the result of inadequate glycaemic control or malformations; (d) twin-to-twin transfusion syndrome in monozygotic twinning; (e) fetal hydrops (rhesus isoimmunisation, rubella, syphilis, toxoplasmosis); (f) chorioangioma of the placenta, circumvallate placenta.

The clinical features include discomfort, dyspnoea, dependent oedema, and supine hypertension. The diagnosis is confirmed by ultrasound in about 1% of pregnancies, based on a four quadrant Amniotic Fluid Index (AFI) greater than the 95th percentile for gestational age. Mild hydramnios (AFI 24.1–39.9 cm) is not clinically significant. Complications of severe hydramnios include premature labour, an unstable lie, prolapsed cord, and placental abruption if fluid is allowed to escape too rapidly, and postpartum haemorrhage. Fetal mortality is as high as 60%. Severe pressure symptoms can be relieved by amniocentesis, but the beneficial effect is transient. Indomethacin can be used to decrease fetal urine output and hence amniotic fluid volume, but after 33 weeks there is a risk of narrowing of the ductus and/or pulmonary hypertension. The incidence of mild and proteinuric pregnancy-induced hypertension is increased in hydramnios when the fetus is normal.

Oligohydramios

In a normal pregnancy at least one pool of amniotic fluid should measure at least 2 cm in two perpendicular planes. The AFI (the sum of the largest vertical fluid pockets in each of the four quadrants) is highly predictive of poor pregnancy outcome. Oligohydramnios (AFI less than the fifth centile for gestational age) may be found with growth retardation of the fetus, prolonged pregnancy (associated with a marked increase in the incidence of meconium aspiration syndrome and perinatal asphyxia), urinary tract obstruction, and renal agenesis (Potter's syndrome, which includes low set ears, epicanthic folds, flattening of the nose, micrognathus). If the condition occurs early in pregnancy (the result of premature rupture of the membranes), there may be amniotic adhesions causing severe fetal deformities, including amputation. When amniotic fluid appears virtually absent, perinatal mortality rates approach 90% owing to pulmonary hypoplasia and abnormal or absent renal function. Investigations include ultrasonography to identify fetal abnormality and fetal karyotyping by amniocentesis, cordocentesis or placental biopsy. Oligohydramnios may cause repetitive umbilical cord compression and may compromise umbilical blood flow in labour. Large volume saline amnioinfusions result in improved fetal outcomes and lower caesarean section rates.

Amniotic Fluid Embolism

This may occur in association with precipitate labour, hypertonic uterine action and hydramnios. The reported incidence ranges from 1 in 3000 to 1 in 80 000 pregnancies, accounting for 4–10% of maternal deaths. Entry of particulate matter (meconium, vernix, lanugo) into the maternal circulation causes acute cor pulmonale and DIC. Management includes securing the airway, correcting shock with crystalloids, vasopressor treatment of acute hypotension and congestive heart failure, and anticipating and promptly treating DIC. The diagnosis can be confirmed only at post-mortem. Eighty-five per cent of patients will die, a quarter of them within 1 h of the acute event. The fetal mortality is 40%.

Intra-amniotic Infection

Clinically evident intra-amniotic infection occurs in 0.5–10% of pregnancies. The incidence is higher after obstetric procedures (cerclage, amniocentesis, umbilical blood sampling or transfusion), but the absolute risk is small. The infection is often polymicrobial, with *Bacteroides*, group B streptococci, *E. coli* and other aerobic Gram-negative rods. Fulminant infection with intact membranes may be caused by *Listeria monocytogenes*. Diagnosis is based on maternal fever, maternal or fetal tachycardia, uterine tenderness, foul-smelling amniotic fluid and leucocytosis. Vigorous intrapartum antibiotic therapy and prompt delivery result in an excellent prognosis for the mother and the term neonate. The combination of prematurity and amnionitis results in higher perinatal mortality rates.

Chapter 12

Premature Labour

Premature labour is defined as the onset of regular painful uterine contractions resulting in effacement and dilatation of the cervix prior to 37 completed weeks of pregnancy. It occurs in 6–8% of pregnancies but is responsible for 75–85% of all perinatal deaths. The aetiology includes elective induction of labour (30%), multiple pregnancy (10%), pregnancy complication, e.g. haemorrhage, infection, cervical incompetence or uterine abnormality (25%), and idiopathic (35%).

A group of patients at increased risk can be identified: black, adolescent, short (less than 150 cm), or small (less than 50 kg). Other factors include high occupational physical exertion, genital tract infection (chorioamnionitis in 50% of preterm and 4% of term labours), a past history of a low birthweight baby, preterm labour (6% recurrence rate), spontaneous, induced or threatened miscarriage.

Management

1. *Prevention*: there is no good evidence that identifying risk factors reduces the incidence of preterm delivery. In multiple pregnancy, bed rest, cervical cerclage or betasympathomimetics are not of convincing benefit in preventing preterm labour.
2. *Tocolytic therapy*: salbutamol (3 mg in 500 ml dextrose/saline at 10 drops/min) or ritodrine (50 μg/min) are increased at the same rate every 15 min until contractions cease or the maternal heart rate exceeds 130 bpm, systolic blood pressure falls by 20 mmHg or the FHR increases by 20 bpm. Beta sympathomimetics inhibit uterine contractions by binding to beta adrenergic receptor sites on myometrial cells, stimulating adenyl cyclase, increasing production of cyclic AMP and decreasing free calcium. Maternal side-effects include tachycardia, hypotension, tremor, nausea, pulmonary oedema and myocardial ischaemia. Fetal side-effects include hypocalcaemia, hypoglycaemia, hypotension and ileus. Contraindications include maternal conditions such as cardiac disease, hyperthyroidism, sickle cell disease and diabetes, and situations in which prolongation of pregnancy may be hazardous such as pre-eclampsia, APH and chorioamnionitis.

Beta sympathomimetics are given only in unexplained preterm labour to patients who otherwise appear totally healthy, have a normal healthy fetus in utero, are no more than 34 weeks' gestation with a fetus weighing no more than 1.5–2.0 kg, have intact membranes and cervix less than 3 cm dilated, i.e. in approximately 10% of cases. They unequivocally reduce the incidence of delivery within 24 h, delivery before 37 weeks and birthweight below 2.5 kg, but this does not result in a detectable decrease in either perinatal mortality rate or the incidence of severe respiratory distress syndrome. The main benefit of tocolytic therapy is a short delay in delivery to allow other measures to be implemented, e.g. in utero transfer to a neonatal intensive care unit or corticosteroid therapy. Other agents such as alcohol, calcium channel blockers and antiprostaglandins have been used to stop premature labour.

3. *Corticosteroids*: corticosteroids (2 mg between 28 and 32 weeks) reduce the incidence and mortality of respiratory distress syndrome, though there is no significant effect on long-term development. The greatest benefit is in infants weighing 750–1250 g, delivered more than 72 h after treatment has begun. Corticosteroids have no long-term adverse effects on infant development, though they may predispose to fetal distress and neonatal hypoglycaemia. They should be used with caution in hypertensive women with a growth-retarded fetus. Maternal leucocytosis occurs but there is no increase in postpartum sepsis.

4. *Labour management*: in utero transfer is easier, cheaper and safer than neonatal transfer and is associated with reduced neonatal morbidity and mortality. Epidural analgesia in labour is ideal. Changes in the FHR pattern carry the same significance in preterm as in term labour as do scalp pH measurements, except that deterioration may occur more rapidly. A quarter of preterm infants are also growth retarded and therefore more likely to become asphyxiated. The mode of delivery is determined by weight and presentation. The following conclusions are based on retrospective studies in which it is impossible to avoid bias. Survival of infants weighing less than 1.0 kg may be better after caesarean section but for the breech presentation it is largely dependent on available facilities. At birthweight between 1 and 1.5 kg, caesarean section makes no difference if the presentation is cephalic. With a breech presentation of between 1 and 1.5 kg, the survival rate following abdominal delivery is three times that of vaginal delivery; if the breech is also growth retarded, the difference is fivefold. Survival in the breech weighing 1.5–2.5 kg is not affected by mode of delivery. Prophylactic forceps and episiotomy make no difference to neonatal mortality or morbidity in low birthweight infants, but if the birthweight is between 0.8 and 1.3 kg, the incidence of intraventricular haemorrhage is twice as high in the group electively delivered by low forceps. Caesarean section is not without risk in preterm labour because the lower segment is poorly defined: there is a 5–10% risk of ruptured uterus in the next pregnancy.

Outcome

Contractions will cease and pregnancy continue in 50% of those in apparent established preterm labour. If the cervix is uneffaced and less than 3 cm dilated in spite of regular, frequent uterine contractions, absent fetal breathing movements may predict progressive labour within 48 h.

Perinatal Mortality and Morbidity

Survival rates improve dramatically from 23 weeks (6%) to 28 weeks (78%) and after 32 weeks are almost equivalent to those at term. Seventy per cent of infants of more than 1 kg and 35% more than 0.7–0.8 kg will survive; 50% of deaths occur antepartum or are due to lethal anomalies. Ten per cent of survivors are handicapped, the severity being inversely proportional to gestational age. Intraventricular haemorrhage occurs in 30–50% of preterm babies weighing less than 1.5 kg at birth; it is multifactorial in origin (respiratory distress syndrome, artificial ventilation, hypercapnia, acidosis and hyperthermia). The incidence is lowest in infants that do not require postnatal transfer. Bronchopulmonary dysplasia is a chronic condition affecting survivors of long-term mechanical ventilation in which supplemental oxygen is needed to maintain an arterial PO_2 of greater than 7 kPa for 28 days or more after birth.

Complications include cerebral palsy, mental retardation, growth deficit and SIDS. Major neurological handicap is a disability that may prevent the child from attending a normal school or seriously interfere with normal function in society. It occurs in 5–8% of low birthweight infants (less than 2.5 kg), 18% of very low birthweight (less than 1.5 kg) and 49% of infants weighing 0.7–0.8 kg.

Preterm Premature Rupture of Membranes

This occurs in one third of preterm labours. The diagnosis is by history and physical examination (amniotic fluid seen trickling from cervix), confirmed by nitrazine sticks, which give a positive diagnosis in 95% of cases. If there is still doubt, giving oral pyridium will stain urine orange. The incidence of infection increases with duration of membrane rupture and decreases with gestational age: there is a 13% risk of infection within 24 h of rupture at 28–29 weeks, 7% at 30–31 weeks and 1% after 35 weeks. Most fetal deaths in preterm premature rupture of membranes are due to prematurity rather than infection. The dilemma of management is to balance the risks of maternal/fetal infection inherent in conservative treatment against the risks of preterm delivery. Conservative management prior to 34 weeks includes avoiding digital examination, a single sterile speculum examination to take high vaginal and cervical swabs for bacteriology, and monitoring maternal temperature, tachycardia, uterine tenderness, vaginal discharge, leucocytosis and fetal tachycardia, all of which are late signs of infection. Prophylactic antibiotics are not helpful. Amniotic fluid may be cultured and the lecithin:sphingomyelin ratio measured. If greater than 2 and especially if organisms are present in the amniotic fluid, delivery is expedited; otherwise, amniocentesis is repeated after 24 h of corticosteroid treatment; the theoretical risk of increased infection rates has not been borne out in clinical practice. The major contraindication to conservative treatment is the presence of herpes virus or group B beta haemolytic streptococcus infection.

Chapter 13

Multiple Pregnancy

The rate of twin pregnancy is 6 in 1000 births in Japan, 10 in 1000 in the UK and 66 in 1000 in Nigeria. Most of this variation is due to different rates of dizygotic (DZ) twinning. The rate of monozygotic (MZ) twinning is fairly constant at 3-5 in 1000. Triplets occur once in every 6400 births and quadruplets once in every 512 000 births.

There are no specific factors known to affect the incidence of MZ twinning. Factors associated with increased frequency of DZ twinning include: (a) mothers aged 35-39 years; (b) increased parity independent of maternal age; (c) tall women; (d) previous twins; (e) family history of twins on the maternal side; (f) good nutrition; and (g) ovulation induction.

Mechanisms of Twinning

Dizygotic (DZ) twinning is duplication of the process of conception as a result of fertilisation of two separate ova. Each child has its own membranes (chorion and amnion; dichorionic, diamniotic) and placenta. Even though the placentas may fuse, the circulations remain separate. Superfecundation is the result of fertilisation of two ova by two separate acts of coitus. Superfetation is DZ twinning arising from coitus in different cycles.

Monozygotic twinning arises from splitting of a single embryo. The result depends on when this occurs:

1. 3 days post-fertilisation (8 cell stage; one third of cases): two normal blastocysts form; the result is the same as a DZ pregnancy except that the twins are identical.
2. 4-7 days (inner cell mass forming; two thirds of cases): there is a single chorion and placenta, with anastomoses between the two circulations; usually there are two separate amniotic membranes and cavities (monochorionic, diamniotic).
3. 8-12 days (1%): a single amniotic cavity forms (monochorionic, monoamniotic).
4. 13 days onwards (very rare): gives rise to conjoined twins (most commonly thoracopagus).

The type of twinning can be assessed by ultrasound examination of the membranes in the first trimester, or by the following at delivery:

1. Fetal sex: if the sexes are different, the twins must be DZ.
2. Examination of membranes: if there is only one chorionic membrane the twins must be MZ.
3. If the membranes are dichorionic the twins may be MZ or DZ.
4. Genetic markers: include blood groups, histocompatibility antigens, DNA "fingerprinting", and finger or palm prints.

Triplets are usually trizygotic. A combination of DZ and MZ is unusual; MZ triplets are very rare.

In the mother, most of the physiological changes associated with a normal pregnancy (respiratory, renal, etc.) are exaggerated in a twin pregnancy. There is a greater than normal increment in red cell mass, but haemoglobin and plasma iron are lower than normal.

Complications of Multiple Pregnancy

The mechanical effects of a multiple pregnancy (backache, constipation, varicose veins etc.), become prominent only in the second half of pregnancy.

Fetal malformations are commoner but only in MZ pregnancies. Both twins may be affected, e.g. concordance of 30% for congenital dislocation of the hip. Chromosome anomalies are no commoner than would be expected for women in the relevant age groups.

Miscarriage, hyperemesis, antepartum haemorrhage (placenta praevia) and anaemia are considered to be commoner in multiple pregnancy, though some studies have been unable to confirm the latter two associations. Polyhydramnios may occur; acute polyhydramnios is more frequent with MZ than with DZ twins. The incidence of pre-eclampsia is three to five times that in singleton pregnancies (both DZ and MZ). The condition tends to occur earlier in pregnancy and to be more severe. A twin pregnancy is much more likely to cause severe rhesus sensitisation. If the mother is rhesus-immunised and the father is heterozygous rhesus-positive, then it is possible for one of a pair of DZ twins to be affected by haemolytic disease while the other is not.

Twin-to-twin transfusion can occur through the placental anastomoses in a MZ pregnancy. There is often polyhydramnios associated with the plethoric recipient twin, and oligohydramnios with the donor; the donor is anaemic and growth retarded. Evidence of this process is found in 15% of MZ pregnancies (difference in haemoglobin concentration of more than 5 g/dl) but mild examples may go unrecognised. The donor twin may die from hypoxia and the recipient from cardiac failure. The diagnosis of this syndrome can be made by ultrasound and Doppler ultrasound during pregnancy, and sometimes by heart rate monitoring during labour. After birth the donor baby may require exchange blood transfusion, whereas the recipient may require exchange plasma transfusion.

At least one third of twin fetuses have birthweights below the tenth centile, and some degree of "growth retardation" is probably almost universal. Unexplained intrauterine death is commoner in MZ twins. Preterm birth is common (30% versus 5–6% in singletons; 10% versus 1% before 32 weeks) and more than half the babies weigh less than 2500 g. Survival rates are no worse than for corresponding singleton birthweights. Preterm labour is commoner with MZ than with DZ twins.

The perinatal death rate in twins is five to six times that in singletons. Morbidity (cerebral palsy, learning difficulties, etc.) are also increased. The rates are higher in second than in first twins, and in MZ than in DZ.

Antenatal Management of Twins

An early scan showing two or more gestation sacs should be confirmed later; 10% of multiple pregnancies are missed by early scans. The diagnosis can be made by physical examination, but twins are more commonly identified by a booking ultrasound examination. The usual sequence of antenatal visits is adequate in many cases, but with close attention being paid to nutrition, avoidance of tiredness, etc. Standard tests should be performed, including biochemical screening for neural tube defects and Down's syndrome, though the results of the latter may be difficult to interpret or act upon because the levels of AFP, hCG and other fetoplacental products are higher than normal. For multiple pregnancy, nuchal translucency may be the best screening test. Amniocentesis for karyotyping may be offered for the usual indications, but requires special skills. If only one child is abnormal, selective fetocide may be considered.

A twin pregnancy should be treated as any high risk pregnancy, with regular ultrasound assessment of fetal growth, CTG and biophysical profile scanning.

The prevention of preterm labour follows principles similar to those in singleton pregnancies, recognising that the multiple pregnancy is at particularly high risk of this complication. The value of hospital bedrest between 32 and 36 weeks in women without complications is no longer favoured.

Management of Labour

In 40% of cases both twins are vertex presentations, and in 70% the first twin is a vertex. The second twin is at greater risk of malpresentation, intrapartum hypoxia and cord prolapse. Mechanical problems are rare because of the relatively small size of the fetuses.

In preterm labour or if one or both twins are in a transverse lie it is probably best to perform caesarean section. With vertex and/or breech presentations vaginal delivery should be encouraged.

The general management of labour is similar to that of any high risk pregnancy. An intravenous line is essential and epidural analgesia is desirable. An anaesthetist should be present at delivery. As soon as the first twin is delivered the first cord is divided to obviate the risk of haemorrhage from the second twin. The second twin should be delivered reasonably soon after the first. The exact interval is not important, provided that the condition of the second twin remains satisfactory. Uterine contractions are encouraged with a continuous infusion of oxytocin prior to rupturing the second sac. Assisted delivery of the second twin may be by ventouse if cephalic, or breech extraction if breech with a complication, e.g. for cord prolapse. The use of emergency caesarean section for the second twin is increasing.

In very rare cases the second twin may deliver successfully several weeks after the first.

The risk of postpartum haemorrhage is minimised by oxytocin infusion continuing throughout the third stage.

Occasionally routine oxytocics are given in the presence of an undiagnosed second twin. Vigorous uterine contraction may then produce fetal hypoxia. The management is immediate delivery, by caesarean section if necessary. Some use tocolytics or general anaesthesia (halothane) to relax the uterus.

Triplets or larger multiple births should be delivered by caesarean section, as should locked or conjoined twins.

Chapter 14

Maternal Diseases in Pregnancy

Cardiovascular Disease

Heart disease affects approximately 1% of all pregnancies, of which the majority are congenital in origin. The incidence of rheumatic heart disease has declined.

Physiological Changes in the Cardiovascular System in Pregnancy

Pregnancy is accompanied by significant haemodynamic changes (Table 14.1) that account for the dyspnoea, tachycardia and reduced exercise tolerance that are common in normal pregnancy; they make the diagnosis of certain forms of heart disease more difficult.

Pregnancy Management in Heart Disease

Ideally, preconception counselling will identify women with serious heart disease, e.g. cardiomyopathy, primary pulmonary hypertension, Eisenmenger's syndrome and Marfan's syndrome, to advise them about the dangers of pregnancy (Table 14.2). Early attendance at a combined antenatal and cardiac clinic allows timely decisions regarding the need for termination or surgery. Echocardiography is safe during pregnancy, but cardiac catheterisation with contrast angiography and CT may pose risks to the fetus.

Open heart surgery is performed for life-threatening pulmonary oedema unresponsive to medical treatment. Closed mitral valvotomy may be indicated in severe mitral stenosis (the most common rheumatic disease in pregnant women), when the increased heart rate, blood volume and transvalvular flow of pregnancy accentuate its severity. The optimal time for surgery is the second trimester, when the haemodynamic burden is least.

Scrupulous attention to endocarditis prophylaxis is critical during pregnancy, as is avoiding risk factors for heart failure, e.g. hypertension, infection, anaemia and arrhythmias. Limiting physical activity, salt restriction and drugs (digoxin,

Table 14.1. Cardiovascular changes occurring in pregnancy

Parameter	Change
1. Haemodynamic	
Cardiac output	40% increase (by 1.5 l/min); declines in last 8 weeks
Pulse rate	15% increase (16 bpm), maintained till term
Stroke volume	30% increase, declines towards term
Plasma volume	40% increase; maximal at 34 weeks, then falls till term
Total peripheral resistance	25% decrease, nadir at 14–24 weeks
Blood pressure:	
a. Systolic	Constant
b. Diastolic	15% decrease by 28 weeks, then normalises by term
2. Clinical features	
Jugular venous pressure	Unchanged
Widely split, loud S1 and audible S3	85% of patients
Ejection systolic or continuous murmur[a]	95% of patients
3. Investigations	
Chest X-ray	Slight cardiomegaly; increased pulmonary vascular markings; pulmonary venous distension
ECG	T wave flattening or reversal in lead III, ST segment depression, lower voltage QRS complexes, deep Q waves, occasional U waves
Echocardiogram	Ventricular wall muscle mass, diastolic volume and ejection fraction increase

[a] A systolic murmur is considered significant if it is of grade > 3/6, varying with respiration or associated with other abnormalities. Diastolic murmurs are considered abnormal during pregnancy.
ECG, electrocardiogram.

Table 14.2. Specific cardiac diseases in pregnancy

Disease	Effect of pregnancy	Investigation/treatment in pregnancy
Congenital heart disease		
Atrial septal defect	None in the absence of pulmonary hypertension; reverse shunting through ASD; rarely paroxysmal atrial flutter	Antibiotic prophylaxis in labour rarely required
Persistent ductus arteriosus	Well tolerated; flow reversal with pulmonary hypertension or systemic hypotension	Antibiotic prophylaxis for delivery; anticoagulants for pulmonary hypertension
Ventricular septal defect	Increased cardiac output with large VSD causes congestive heart failure; arterial hypotension causes shunt reversal; the higher the pulmonary resistance, the greater the maternal risk	Cardiac catheterisation when pulmonary hypertension suspected; consider TOP for severe pulmonary hypertension; antibiotic prophylaxis
Pulmonary stenosis	Generally well tolerated if mild/moderate	Antibiotic prophylaxis
Eisenmenger's syndrome	Aggravates condition: 50% risk of maternal death; if cyanotic, fetal mortality > 50%	Avoid pregnancy; limit physical activity, low flow oxygen, planned term delivery with haemodynamic monitoring
Fallot's tetralogy	Well tolerated unless cyanotic	Antibiotic prophylaxis; surgical correction prior to pregnancy

Maternal Diseases in Pregnancy

Table 14.2. (continued)

Disease	Effect of pregnancy	Investigation/treatment in pregnancy
Coarctation of aorta	Hypertension may cause rupture of associated berry aneurysm or aorta	Consider TOP for unoperated patients in heart failure, history of CVA or uncontrolled hypertension; antibiotic prophylaxis
Aortic stenosis	Chest pain, syncope and sudden death are rare; dilated ventricle is ominous	Echocardiography to determine severity; valve replacement after pregnancy; avoid vasodilators; antibiotic prophylaxis
Hypertrophic obstructive cardiomyopathy	Generally well tolerated; inferior vena cava compression causes hypotension and syncope	Beta blockers if symptomatic; avoid digitalis, hypovolaemia, supine position; oxytocin is safe; antibiotic prophylaxis
Congenital heart block	Usually able to increase cardiac output whether paced or not	Pacing may be required
Rheumatic heart disease		
Mitral stenosis	Aggravated severe MS: pulmonary oedema	Digoxin, beta or calcium channel blocker or mitral valvotomy prior to pregnancy for AF; if persistent may require anticoagulation; antibiotic and digoxin prophylaxis; avoid supine position
Mitral regurgitation	Heart failure rare; endocarditis; risks increased if MR severe	Reduced physical activity, salt restriction, digoxin and diuretics; antibiotic prophylaxis
Aortic regurgitation	Well tolerated unless LVF is present	Antibiotic prophylaxis
Artificial valve replacement	Usually well tolerated	Warfarin, then heparin after 37 weeks improves outcome; antibiotic prophylaxis
Pregnancy cardiomyopathy	RVF in puerperium usually in multiparous black women > 30 years	Digoxin, diuretics, anticoagulants; avoid repeat pregnancy unless transplanted

ASD, atrial septal defect; VSD, ventricular septal defect; TOP, termination of pregnancy; CVA, cardiovascular accident; MS, mitral stenosis; AF, atrial fibrillation; MR, mitral regurgitation; LVF, left ventricular failure; RVF, right ventricular failure.

diuretics and vasodilators, but not angiotensin-converting enzyme (ACE) inhibitors) may be required to treat heart failure in pregnancy.

Anticoagulants are indicated in patients with pulmonary hypertension, artificial valve replacements and atrial fibrillation to prevent systemic thromboembolism. In spite of the fact that up to 30% of warfarin-treated pregnancies (Food and Drug Administration (FDA) category X: risks outweigh potential benefits) have adverse outcomes (including spontaneous miscarriage, preterm delivery, fetal or neonatal death), with a teratogenic risk of up to 10% in infants born alive, warfarin gives better anticoagulant protection than heparin; it should be used until 37 weeks and replaced with heparin to reduce the risk of fetal bleeding in labour. Warfarin is recommended a week after delivery. If necessary, intramuscular vitamin K may be given to reverse the anticoagulant effects of warfarin in the fetus. Breast feeding is not contraindicated.

Beta blockers may cause neonatal respiratory depression, sustained bradycardia and hypoglycaemia when given late in pregnancy. ACE inhibitors are contraindicated in pregnancy (fetal and neonatal renal complications). Neonatal electrolyte imbalance, jaundice, thrombocytopenia, liver damage and death may follow thiazide use.

Labour is generally rapid and uncomplicated in women with cardiac disease. The main risks are fluid overload (pulmonary oedema) and aortocaval compression (hypotension). Epidural anaesthesia is safe except in Eisenmenger's syndrome, where by decreasing systemic vascular resistance it could further decrease pulmonary blood flow, and in hypertrophic cardiomyopathy, where by increasing lower limb venous capacitance it causes relative hypovolaemia. Elective caesarean section under general or epidural anaesthesia is advocated by some for severe heart disease. The second stage may be shortened by a forceps delivery. Intravenous oxytocin may be combined with frusemide in the third stage of labour to reduce the risk of postpartum haemorrhage. Antibiotic prophylaxis is used in labour to prevent acute bacterial endocarditis. Beta sympathomimetics cause tachycardia and vasodilatation, both of which increase the risk of pulmonary oedema; they should be given only with great care to women with heart disease, especially mitral stenosis.

Maternal and Fetal Prognosis

This is now greatly improved. Heart disease causes approximately 10 maternal deaths per million maternities in England and Wales. Mortality is higher (up to 50%) when pulmonary blood flow cannot be increased because of obstruction, e.g. Eisenmenger's syndrome and primary pulmonary hypertension. Pregnancy does not affect the long-term survival of women with rheumatic heart disease, provided they are among the 99% of women who survive their pregnancy. Pulmonary oedema is the major cause of death.

Fetal outcome is generally good except in women with uncorrected cyanotic congenital heart disease or open heart surgery in the first trimester. Congenital heart disease in the mother may recur in the fetus, the risk ranging from 5% in atrial septal defects to 25% in Fallot's tetralogy.

Respiratory System

The pregnant woman compensates for increased oxygen consumption with a progesterone-driven increase in tidal volume. This leads to maternal hyperventilation and hypocapnia. Anatomical and mechanical changes modify acid–base balance such that respiratory alkalosis is normal in pregnancy (Table 14.3). Breathlessness occurs in 50% of pregnant women, mostly at the beginning of the third trimester.

Asthma

Asthma affects 3% of women of reproductive age and is the commonest respiratory disorder complicating pregnancy. Pregnancy has no consistent effect on

Table 14.3. Changes in respiratory system during pregnancy

Parameter	Value	Change	Result
Diaphragm		Level rises, movement increases	Dyspnoea
Ventilation rate	10.5 l/min	40% increase	
Tidal volume	700 ml (T3)	200 ml increase	Compensated respiratory alkalosis
Residual volume	1 l (T1)	200 ml decrease	Compensated respiratory alkalosis
Vital capacity	3.2 l	Unchanged	
O_2 consumption	300 ml/min	15% increase	
PA_{CO_2}	4 kPa or lower	Decreased	Renal compensation
HCO_3^-	18–21 meq/l	Decreased	

T1, first trimester; T3, third trimester

asthma. Exacerbations during labour are rare. Individual patients tend to have a similar course during successive pregnancies. If adequately controlled, asthma does not increase perinatal morbidity or mortality. Severe asthmatics on oral steroid therapy have a slightly higher incidence of intrauterine growth retardation and prematurity. Cleft palate has been reported in animal but not human studies.

Management in Pregnancy

As in the non-pregnant patient, peak flow monitoring at home allows the patient to alter treatment before symptoms deteriorate. Treatment is with regular betamethasone inhalations; beta-2 sympathomimetic inhaler (salbutamol, terbutaline or albuterol), and oral slow release aminophylline are reserved for breakthrough symptoms. A short course of oral prednisolone may also be required. If the pulse rate exceeds 100, the respiratory rate is 30 or more per minute and the peak expiratory flow rate is less than 100 l/min, hospital admission for oxygen inhalation (10 l/min via mask) is warranted. Mechanical ventilation may be required if the Pa_{CO_2} starts to rise.

Management of Labour

If necessary, PG E2 rather than F2α should be given for cervical ripening, because it is a bronchodilator. Hydrocortisone may be required to reduce the risk of suppression of the maternal hypothalamo–pituitary–adrenal axis. Fetal suppression leading to neonatal collapse is rare because (a) very little prednisone crosses the placenta, (b) placental 11β-dehydrogenase effectively deactivates prednisone, and (c) the fetoplacental unit is relatively lacking in the enzymes necessary to convert prednisone into its active metabolite prednisolone. Epidural anaesthesia is preferable, but, if general anaesthesia cannot be avoided, the bronchodilatating effect of halothane may help. Pre-and postoperative salbutamol is given. Breast feeding is not contraindicated.

Respiratory Drugs in Pregnancy

Treatment with steroids, beta sympathomimetics or theophylline is unlikely to affect the fetus. The maternal side-effects of beta sympathomimetics include tachycardia, tremor and anxiety. There is no evidence of teratogenicity. Rarely, high dose intravenous therapy may cause pulmonary oedema and metabolic acidosis. These drugs cause hyperglycaemia: they are best avoided in patients with impaired glucose tolerance. They do not prolong pregnancy or impede the progress of normal labour. Aminophylline is safe in pregnancy; very high intravenous doses may cause fetal jitteriness but there are no long-term sequelae. Prophylactic disodium cromoglycate is safe for the fetus. At a dose of less than 20 puffs (1 mg), inhaled beclomethasone does not enter the blood. Long-term therapy causes monilial infections in fewer than 5% of users. Tetracycline can cause permanent staining of the infant's teeth and bone abnormalities and should be avoided in the treatment of chest infection. Iodide-containing expectorants should be avoided as the iodine is actively taken up by the fetal thyroid, causing goitre or hypothyroidism.

Infections

Viral upper respiratory tract infections occur as in non-pregnant women; symptomatic treatment suffices. Acute bacterial infections causing bronchitis may require antibiotics; co-trimoxazole (only theoretical risk), aminoglycosides and tetracycline are contraindicated, as are iodine-containing expectorants. Pneumonia presents with cough, purulent sputum, chest pain and pyrexia, typically in smokers. Differential diagnosis from pulmonary embolism is aided by low central venous pressure in infection. Most infections are bacterial and respond to the appropriate antibiotic. Premature labour may be precipitated by pyrexia.

Pulmonary Complications of Human Immunodeficiency Virus (HIV)

Screening for HIV during pregnancy, with pre- and post-test counselling is recommended in the USA and high risk areas in the UK, e.g. London. HIV-infected pregnant women should receive Pneumovax, influenza and hepatitis vaccines, and should be screened for tuberculosis and sexually transmitted diseases. Immune status is monitored with CD4 counts; if the count drops below 200/ml, *Pneumocystis carinii* pneumonia (PCP) prophylaxis (co-trimoxazole or nebulized pentamidine) is started. PCP is associated with a high incidence of adverse obstetric outcome.

Pregnancy has no consistent adverse effect on the course of HIV disease and vice versa. Transplacental transmission, the major route of infection among infants is prevented by azidothymidine through its inhibiting effect on reverse transcriptase. Caesarean section does not prevent infection in the child. Where safe alternatives to nursing are available, mothers known to have HIV disease should avoid breast feeding.

Tuberculosis

Tuberculosis rarely complicates pregnancy in the UK except in certain ethnic groups or HIV-positive women. The diagnosis is the same as in non-pregnant patients. Mantoux status is not affected by pregnancy. Transplacental passage of the bacillus is rare.

Ethambutol is safe in pregnancy. If isoniazid is used it should be combined with pyridoxine to reduce the risk of peripheral neuritis. Once organogenesis is complete, rifampicin is added. Treatment is continued for 9 months. Streptomycin affects the VIIIth cranial nerve causing deafness: it is contraindicated in pregnancy. Multidrug-resistant strains of tuberculosis are emerging in HIV-positive individuals.

Sarcoidosis

Sarcoidosis affects 0.05% of all pregnancies. It generally improves slightly in pregnancy, with a tendency to relapse in the puerperium. Women presenting only with symmetrical, bilateral hilar lymphadenopathy are usually asymptomatic and do not warrant treatment. Steroid therapy is required if systemic symptoms are severe, diffuse lung infiltration or other organ involvement are present, or if there is hypercalcaemia. Patients should not take extra vitamin D, which may cause hypercalcaemia.

Kyphoscoliosis

Kyphoscoliosis is found in 0.1–0.7% of all pregnant women. Operative delivery is more likely to be required because of associated bony pelvis abnormalities. In spite of the spinal deformities, epidural anaesthesia is preferable. Pregnancy outcome is usually successful even when the vital capacity is less than 1 litre but perinatal mortality is increased in hypoxic women with pulmonary hypertension.

Cystic Fibrosis

Cystic fibrosis is an autosomal recessive condition causing airflow obstruction, pulmonary restriction or both; it can now be managed by gene therapy. As many as half the patients have been reported to deteriorate during pregnancy.

Bronchial toilet and treatment of chest infections should continue during pregnancy. Both dehydration and overhydration, which may precipitate right-sided cardiac failure, should be avoided in labour. Epidural anaesthesia is preferable. Breast feeding is not contraindicated if the sodium content of the milk is normal. There is a higher perinatal mortality rate than in the general population, but maternal mortality is no higher than in non-pregnant cystic fibrosis patients.

About 4% of the white population carries the cystic fibrosis gene. One in 40 offspring of affected individuals have cystic fibrosis; all other infants are obligate heterozygotes. Because of the possibility of gene mutations, most heterozygous carriers, but not all, can be identified from blood or buccal cells. First trimester prenatal diagnosis on chorionic villus samples is then possible after genetic counselling.

Adult Respiratory Distress Syndrome

Adult respiratory distress syndrome may be precipitated by: (a) inhalation of gastric contents during anaesthesia; (b) DIC secondary to pre-eclampsia, eclampsia, abruption, sepsis, dead fetus syndrome or amniotic fluid embolism; (c) hypovolaemic shock; (d) severe anaphylaxis; and (e) hydatidiform mole.

Severe acute hypoxaemia in spite of adequate inspired oxygen concentration is characteristic. Chest radiograph shows diffuse pulmonary infiltrates. Normal pulmonary arterial wedge pressure measurements exclude a primary cardiac abnormality. Management includes correcting the underlying cause, mechanical ventilation, maintaining fluid and electrolyte balance as well as steroids, diuretics and antibiotics. The mortality is in excess of 50%.

Prevention of gastric aspiration is by starving the labouring woman and administering regular antacids (sodium citrate), H2 receptor antagonists (cimetidine) and metoclopramide (increases gastric emptying).

Neurological Disease

Apart from epilepsy and mononeuropathies, neurological diseases are rare in pregnancy (Table 14.4).

Epilepsy

Epilepsy occurs in 1 in 2000 women of childbearing age per year. Certain anticonvulsants interfere with the effectiveness of oral contraceptives. Prepregnancy counselling of known epileptics is essential. The risk to the pregnancy may be higher if anticonvulsants are stopped.

Effect of Pregnancy on Epilepsy

The frequency of seizures is unchanged in approximately 50%, increased in 35% and decreased in 15% of pregnancies. Loss of seizure control in pregnancy is most likely to be due to poor compliance resulting from fear of teratogenesis or nausea, sleep deprivation and altered phamacokinetics of anticonvulsants. The dose of anticonvulsant often needs to be increased during pregnancy to maintain plasma concentration at a level previously known to be effective. Plasma levels do not reflect the free (biologically active) drug concentration; monitoring of blood anticonvulsant levels is of little value unless serum binding capacity is also measured.

Only rarely does pregnancy itself initiate epilepsy. Less than a quarter of such cases will have recurrent seizures only in subsequent pregnancies (gestational epilepsy). The differential diagnosis of seizures occurring during labour or the puerperium includes eclampsia (with hypertension and proteinuria). Status epilepticus does not occur more frequently during pregnancy. When it does occur the maternal and fetal mortality is high. Treatment is with intravenous diazepam and anticonvulsants.

Effect of Epilepsy (and its Treatment) on Pregnancy

There is an increased stillbirth and neonatal death rate, especially if seizures occur during pregnancy. The incidence of congenital abnormalities, most commonly facial clefts and congenital heart disease, is twice that in the general population. Both epilepsy and its treatment increase the risk, especially when anticonvulsants are combined. The fetal hydantoin syndrome includes small digits, midline fusion defects, microcephaly, mental retardation, congenital heart disease and hypertelorism. Sodium valproate is associated with an increased incidence of neural tube defects. Prenatal diagnosis is possible. Phenytoin may induce

Table 14.4. Clinical features of neurological diseases in pregnancy

Clinical features and incidence in pregnancy	Effect of/on pregnancy	Investigations/treatment
Cerebrovascular disease		
Arterial occlusion 1:20 000 live births	Mortality three times that of non-pregnant women	CT; steroids
Cerebral venous thrombosis	Haemodynamic changes predispose to thrombosis especially in puerperium	Cerebral arteriography and CT scan: if no intracranial haemorrhage treat with heparin; steroids and anticonvulsants
Subarachnoid haemorrhage	More likely to bleed during pregnancy or labour. High maternal and fetal mortality	Angiography and neurosurgery as soon as possible
Neuromuscular disorders		
Myasthenia gravis 1:30 000	40% worse, 30% better and 30% unchanged in pregnancy; those who deteriorate do so postpartum with respiratory failure. 10% of infants have transient neonatal myasthenia; 80% require anticholinesterases. 4% maternal and 8% fetal mortality rate	Symptoms relieved by thymectomy, parenteral neostigmine or pyridostigmine in labour; steroids and/or plasmapheresis. Most labour spontaneously; epidural effective: minimise maternal effort with low forceps delivery
Multiple sclerosis 5/100 000 women of childbearing age. Demyelinating disease	Number and severity of relapses decrease during pregnancy and increase in puerperium; no deleterious effect of pregnancy on long-term prognosis of MS; no increase in congenital abnormality rate; epidural not contraindicated	Cerebrospinal fluid immunoglobulin-evoked response tests, short course of steroids, immunosuppressives and plasmapheresis
Migraine		
Vascular headaches	Symptoms improve in 80%, usually after TI; metoclopramide for acute attacks; ergotamine contraindicated because of oxytoxic effects, also in lactating women	Avoid precipitating factors; aspirin and metoclopramide for acute attacks; ergotamine contraindicated because of oxytoxic effects, also in lactating women
Benign intracranial hypertension Raised intracranial pressure of unknown aetiology	Occurs more frequently in pregnancy but of shorter duration; usually self-limiting; recurrence is rare; maternal and neonatal outcome good	Exclude other causes with CT scan; monitor visual acuity and fields; steroids and diuretics; epidural preferable, mode of delivery unaffected
Paraplegia	Urinary tract infections are major risk; labour painless if cord transection complete above T 10; forceps delivery often required	As in non-pregnant
Nutritional disorders (Wernicke's encephalopathy) ataxia, nystagmus, diplopia, altered consciousness	Increased thiamine requirement in pregnancy; severe vomiting or hypomagnesaemia	Thiamine

CT, computed tomography; MS, multiple sclerosis; T1, first trimester.

folic acid deficiency by impairing gastrointestinal absorption and/or by increasing hepatic metabolism. Folic acid supplements reduce the risk of megaloblastic anaemia.

By inducing fetal hepatic microsomal enzymes, anticonvulsants decrease the level of vitamin K-dependent coagulation factors in the neonate. Bleeding, usually within 24 h of birth, is prevented by giving the mother vitamin K for 2 weeks before delivery or the infant may be given vitamin K intramuscularly at birth.

Anticonvulsants may cause neonatal intoxication, which presents with hypotonia, hypothermia, reluctance to feed and poor weight gain. Neonatal barbiturate withdrawal results in restlessness, tremors and hyperreflexia. The half-life of anticonvulsants (phenobarbitone, over 100 h; phenytoin, 60 h; sodium valproate: 40 h) influences their clearance rate from the infant's circulation. Breast feeding is encouraged except when sedation is a problem.

The risk of epilepsy in the child of an epileptic mother is increased 10-fold (to 2–5%) compared with the general population. If both parents are epileptic, the risk increases to 15%.

Neuropathies

Mononeuropathies including the carpal tunnel syndrome (median nerve), meralgia paraesthetica (lateral cutaneous nerve of the thigh) and Bell's palsy (facial nerve) are all more common in pregnancy owing to fluid retention. They present with pain, weakness, and paresthesias in the distribution of the nerve. Treatment options in carpal tunnel syndrome include splinting the wrist, hydrocortisone or operative decompression. Symptoms developing or worsening during pregnancy generally resolve within 3 months of delivery, though they may recur in subsequent pregnancies. A short course of oral steroids may be effective if given within 24 h of onset of Bell's palsy. Surgical decompression is often disappointing in lower limb pressure palsies; they frequently remit after delivery.

Polyneuropathies including Guillain–Barré syndrome are rare in pregnancy but carry a high maternal and fetal mortality rate, especially if mechanical ventilation is required. Pregnancy can be allowed to continue to term in mild cases and the infants are not affected by the disease.

Genitourinary Disease

Renal Adaptations to Normal Pregnancy

Dilatation of the renal tract (more on the right than the left) and a 1 cm increase in kidney length occur in pregnancy. This physiological hydronephrosis and hydroureter, which lead to an increase in ascending urinary tract infection related to stasis, may persist for up to 4 months post partum. Other functional changes are shown in Table 14.5. Increased urinary protein excretion implies that proteinuria is not abnormal in pregnancy until it exceeds 300 mg per 24 h. Glucose can spill into the urine at lower blood levels in pregnancy because of decreased tubular glucose reabsorption. All components of the renin–angiotensin–aldosterone system increase in the first trimester. The combination of normal urine volume, lowered plasma osmolarity and raised extracellular fluid volume is unique to pregnancy.

Maternal Diseases in Pregnancy

Table 14.5. Functional changes in the renal tract during pregnancy

Parameter	Change	Average value in pregnancy
Glomerular filtration rate	50% increase in T1; 15% decrease in T3	170 ml/min
Creatinine clearance	20–80% increase in T2, falls to NP value in T3	150–200 ml/min
Renal plasma flow	Declines slowly from 30 weeks	
Serum creatinine	20% decrease	NP = 73, T1 = 65, T2 = 51 μmol/l
Plasma urea	Decreased	T1 = 3.5, T2 = 3.3, T3 = 3.1 mmol/l
Serum sodium	Decreased	NP = 140, T2 = 137 mmol/l
Serum potassium	Decreased, rising to NP values in T3	NP = 4.3, T2 = 4.0 mmol/l
Total body water	Increased by 6–8 litres	
Plasma volume	Increased by 50%, peak in T2, falls by 200–300 ml in T3	
Plasma osmolality	Decreased by 10 mosmol/kg H_2O	280 mosmol/kg H_2O
Urinary frequency	Increased at night	
Bladder tone	Decreased	
Stress incontinence	Increased	

T1, first trimester; T2, second trimester; T3, third trimester; NP, non-pregnant.

Urinary Tract Infection (UTI)

About 4% of pregnant women have asymptomatic bacteriuria (more than 100 000 bacteria per ml of urine). Urinary stasis predisposes to acute symptomatic pyelonephritis (30–40% risk) but not other pregnancy complications. Women with a previous UTI and asymptomatic bacteriuria are at ten times greater risk of UTI in pregnancy. Screening with a mid-stream urine specimen identifies these women. Asymptomatic bacteriuria is associated with structural abnormalities in the renal tract in 20% of cases. These are identified at intravenous urography at least 4 months after delivery. Persistent asymptomatic bacteriuria per se does not cause long-term renal damage.

Acute infection occurs in about 2% of women during pregnancy, causing dysuria, frequency, loin pain and pyrexia. Symptomatic infections may cause preterm labour and increased perinatal loss. Most initial and recurrent infections are caused by *E. coli*. Gram-negative infection is associated with instrumentation of an infected urinary tract. Ampicillin and cephalosporins are safe in pregnancy; tetracyclines are contraindicated; nitrofurantoin and sulphonamides should be avoided in late pregnancy as they may cause neonatal haemolytic anaemia and hyperbilirubinaemia, respectively. Urinary bacteriology is repeated at every subsequent antenatal visit.

Chronic Renal Disease

The diagnosis is generally known prior to pregnancy (Table 14.6). The risk of pyelonephritis is increased 10-fold (20%) and a quarter will develop hypertension during pregnancy. Distinction from pre-eclampsia may be difficult. A past or family history of renal disease, or more than one red or two white cells per high power field with casts, suggests pre-existing renal disease, but this may

Table 14.6. Clinical features of chronic renal disease in pregnancy

Disease	Effect of/on pregnancy	Investigation/treatment in pregnancy
Diabetic nephropathy	No permanent worsening of renal function. Dramatic increase in proteinuria in T3. Peripheral oedema, pre-eclampsia and bacteriuria more frequent	Careful glycaemic and BP control; early delivery for growth retardation, fetal compromise, substantial deterioration in renal function, uncontrollable hypertension
Renal calculi	Prevalence in pregnancy 1 in 300 to 1 in 3000. More frequent UTIs. Unaffected by pregnancy	50% of stones pass spontaneously; IVP prior to surgery if renal function deteriorating. Adequate hydration, antibiotics, analgesia
Chronic glomerulonephritis	If hypertensive, high incidence of spontaneous miscarriage, prematurity and fetal loss; otherwise no adverse effects except UTIs more frequent	Monthly urine cultures; antihypertensives if required; intensive fetal surveillance (ultrasound, kick chart, CTG, Doppler)
Chronic pyelonephritis	Normotensive patients with good renal function have best prognosis. Perinatal mortality 10% with hypertension or pre-eclampsia	Keep rested and well hydrated; left lateral position is best
Lupus nephropathy	Course unaffected by pregnancy; fetal wastage higher in active SLE (renal or otherwise); best prognosis if in remission for 6 months before pregnancy	Active disease requires steroids; increased dosage in labour and puerperium; early delivery or TOP does not alleviate progressive renal dysfunction
Periarteritis nodosa and scleroderma	Poor maternal and fetal outcome when associated with hypertension	Terminate pregnancy if renal dysfunction deteriorates
Polycystic kidneys	Often undetected; unaffected by pregnancy; good prognosis if normotensive, with minimal functional impairment	Autosomal dominant inheritance; prenatal diagnosis with ultrasound
Unilateral kidney (congenital absence, hypoplasia, nephrectomy)	Pregnancy generally well tolerated	Antibiotics for infection
Ectopic kidney (usually pelvic)	Other associated urogenital tract anomalies may cause dystocia in labour; more vulnerable to infection	Antibiotics for infection

BP, blood pressure; T3, third trimester; UTI, urinary tract infection; IVP, intravenous pyelogram; CTG, cardiotocography; SLE, systemic lupus erythematosus; TOP, termination of pregnancy.

coexist with pre-eclampsia. Renal biopsy is rarely performed in pregnancy because of the risk of bleeding. Labour may need to be induced if uncontrollable hypertension or risk to the fetus supervene. Perinatal mortality due to preterm labour and intrauterine growth retardation are higher than in uncomplicated pregnancies but pregnancy outcome is generally favourable in the absence of hypertension.

Renal function (as reflected in creatinine clearance) usually improves during pregnancy. If it deteriorates, UTI, obstruction, dehydration or electrolyte imbalance should be excluded. Proteinuria begins or worsens during pregnancy in 50% of cases; of itself it does not signal worsening renal disease or pre-eclampsia.

Pregnancy Management in Chronic Renal Disease

Prepregnancy counselling and combined antenatal and renal care may reduce maternal and perinatal morbidity and mortality. Dietary protein restriction is contraindicated in pregnancy. However, if creatinine clearance is markedly reduced, phosphate, potassium and magnesium intake should be reduced. Monthly urine cultures and prompt antibiotic treatment of infection are mandatory. Deterioration of renal function, development of hypertension or evidence of fetal compromise should prompt hospitalisation for intensive maternal and fetal surveillance, antihypertensive therapy and/or delivery. Renal function may decrease by 15% near term, affecting creatinine levels only minimally. Pregnancy should be allowed to continue in the absence of hypertension, even if proteinuria persists. Delivery in a high risk unit is indicated after 38 weeks.

In the absence of overt renal insufficiency or significant hypertension prior to conception, pregnancy does not accelerate chronic renal disease or affect its long-term prognosis. Though fetal prognosis is less favourable than in healthy women, termination is rarely justified on medical grounds.

Pregnancy in Haemodialysed Patients

Haemodialysis is problematic during pregnancy because: (a) early diagnosis of pregnancy is difficult; (b) spontaneous miscarriage is more common; (c) anaemia may require repeated blood transfusions; and (d) fluctuations in blood pressure may occur, e.g. hypotension during dialysis or hypertension exacerbated by transfusion.

Management includes increasing the length and frequency of dialysis to maintain plasma urea below 20 mmol/l and dietary control. Intensive fetal surveillance and delivery by elective caesarean section is advocated by some.

Pregnancy in Transplant Patients

Renal transplantation restores reproductive function in most patients with chronic renal disease. Problems in pregnancy include: (a) first trimester complications; (b) reversible deterioration of renal function; (c) pre-eclampsia in one third of patients; (d) increased risk of urinary and pulmonary infections especially in the immunosuppressed; (e) allograft rejection (fever, oliguria, renal enlargement, tenderness and deteriorating renal function), though this is no commoner than expected in non-pregnant transplant patients; (f) preterm labour in 50% and small-for-gestational-age infants in 20% of cases; and (g) one third of neonates will have severe complications such as respiratory distress syndrome, adrenocortical insufficiency, septicaemia, hyperviscosity and seizures.

Immunosuppressive therapy is continued in pregnancy. Increased steroid cover and prophylactic antibiotics are given in labour. Vaginal delivery is possible but delivery is often by caesarean section. Breast feeding is possible. Pregnancy has no effect on graft function or survival, but immunosuppressed women have a 35-fold increased risk of developing malignancy over the general population. Contraceptive advice is essential.

Acute Renal Failure

In pregnancy acute renal failure may be due to:

1. *Pre-eclampsia or eclampsia.*
2. *Volume contraction* secondary to prolonged vomiting in hyperemesis gravidarum and concealed or overt antepartum haemorrhage.
3. *Coagulation disorders.*
4. *Septic miscarriage.*
5. *Pyelonephritis*: treatment is with acute dialysis and broad spectrum antibiotics.
6. *Acute fatty liver of pregnancy*: management includes early delivery or termination of pregnancy with supportive treatment of hepatic and renal failure. The prognosis is generally poor.
7. *Idiopathic postpartum renal failure* (haemolytic–uraemic syndrome): acute renal failure with severe hypertension, uraemia, microangiopathic haemolytic anaemia and platelet consumption may occur 3–6 weeks after an uncomplicated pregnancy. The clinical course is fulminant and most patients require dialysis. Prognosis is poor (60% mortality or irreversible renal damage) in spite of intensive treatment.

Musculoskeletal Disease

Physiological adaptation to pregnancy

Backache and a waddling gait are common in late pregnancy. Increased mobility in the sacrococcygeal, pubic and sacroiliac joints may result in pain, tenderness and rarely difficulty with walking (pelvic arthropathy of pregnancy). These pelvic changes revert to normal after delivery. The puerperal osteophytes of Rokitansky are an irregular deposit of bone under the periosteum.

Increased fetal demand on maternal calcium stores is maximal in the third trimester and continues after birth if lactation is established. Activation of renal 1-α-hydroxylase increases 1,25-dihydroxycholecalciferol levels, doubling calcium absorption from the maternal gastrointestinal tract. As a result of binding protein changes, total serum calcium and phosphorus fall in pregnancy. Vitamin D levels are generally unchanged. Elevated calcitonin levels during pregnancy and lactation protect the maternal skeleton from demineralisation.

Systemic Lupus Erythematosus (SLE)

This multisystem autoimmune disease has a prevalence of 1 in 700 women aged 15 to 64 years. Black women have an excess risk five times that of white women. Characteristic features include a malar rash, discoid lupus, photosensitivity, oral ulcers, arthritis, pleurisy or pericarditis, psychosis or seizures, haemolytic anaemia, leucopenia or thrombocytopenia, proteinuria or cellular casts, and elevated titres of anti-DNA antibody, anti-smooth muscle antibody, or anti-nuclear antibody titres.

If active disease is severe, exacerbations are more common in pregnancy especially in the puerperium. Spontaneous or late recurrent miscarriage, hypertension, preterm labour and growth retardation are more common but are unrelated to the severity of the disease. Fetal and neonatal outcome are not related to maternal antibody levels.

The neonatal lupus syndrome includes complete heart block, which is usually permanent, and haematologic abnormalities, and/or discoid skin lesions, which usually resolve. One in three neonates with heart block are born to mothers who have or will develop a connective tissue disease. About 60% of mothers of affected children have an anti-ribonuclear antibody (anti-Ro/SSA).

Prepregnancy counselling to plan pregnancy during remission is ideal. The disease process does not improve and may even worsen after termination of pregnancy. Minor symptoms are relieved with paracetamol (acetaminophen). Falling complement levels, falling creatinine clearance or worsening proteinuria are indications for steroid treatment. Parenteral steroid cover should be given in labour, and the dose reduced only slowly after delivery. Azathioprine may be necessary in life-threatening disease. Women with uncomplicated pregnancies can be allowed to labour spontaneously at term; the remainder are managed as high risk cases.

Most maternal deaths are due to pulmonary haemorrhage or lupus pneumonitis and occur in the puerperium. Pregnancy does not affect the long-term prognosis of the disease. The disease does not always follow the same course in successive pregnancies.

The lupus anticoagulant (also present in patients without SLE) and anticardiolipin antibody are associated with the cardiolipin syndrome: thrombotic episodes, pulmonary hypertension, spontaneous (also recurrent) miscarriage and fetal distress or death. Some authorities advocate prednisone and/or aspirin in pregnant women with high titres of these antibodies. Aspirin is relatively safe in pregnancy except in the few days prior to delivery when clotting defects and premature closure of the ductus arteriosus (leading to pulmonary hypertension) may occur.

Rheumatoid Arthritis

Symmetric small joint arthritis is characteristic. Symptoms improve in most patients, especially in the first trimester, but generally recur after delivery. Limitation of hip and/or neck movement may cause problems in labour or general anaesthesia. Pregnancy has no lasting adverse effect on the disease. Neonatal congenital heart block may occur. Management is with rest, local heat and physical therapy. Gold, antimalarials, and cytotoxic agents are contraindicated in pregnancy. If acetaminophen, aspirin and intra-articular corticosteroids are insufficient, systemic steroids or penicillamine should be used. Help with child care should be anticipated.

Scleroderma

Progressive systemic sclerosis is rare in pregnancy. Termination of pregnancy should be considered for severe renal disease, pulmonary hypertension or myocardial fibrosis. In addition, there is a high risk of stillbirth, preterm labour, perinatal and maternal mortality. Milder cases seem not to affect pregnancy. Transient areas of sclerosis may occur in the newborn.

Osteomalacia and Rickets

Defective bone mineralisation may be due to 1,25-dihydroxycholecalciferol deficiency (type I). Causes include: (a) nutritional deficiency, especially in vegetarians; (b) lack of sunlight; (c) malabsorption secondary to inflammatory bowel disease, jejunoileal by-pass surgery or coeliac disease; (d) diminished synthesis in liver disease (especially cholestatic); (e) chronic renal failure; and (f) drugs, e.g. anticonvulsants, heparin and steroids. Low serum calcium and phosphate, and high alkaline phosphatase are found. Pregnant Asians in the UK, particularly if vegetarian, are at increased risk and should take vitamin D. Neonatal morbidity includes rickets, hypocalcaemia, impaired first year growth and tooth enamel defects.

Women with rickets should take vitamin D prophylactically; they often require caesarean section for delivery because of altered pelvic shape and size. Patients with inflammatory bowel disease and cholestatic liver disease need very high doses of vitamin D because so little of it is absorbed.

Type II osteomalacia results from hypophosphataemia due to (a) Fanconi's syndrome, (b) renal tubular acidosis, or (c) primary hypophosphataemia, familial or acquired.

Primary Hyperparathyroidism

Most cases of primary hyperparathyoidism in pregnancy are due to benign adenomas or parathyroid hyperplasia. Hypercalcaemia may be overlooked unless the total serum calcium is corrected for hypoalbuminaemia. Most women with hyperparathyroidism tolerate pregnancy well and hypercalcaemic crises are rare, but they have a higher perinatal mortality rate. Active placental transfer of calcium results in elevated fetal calcium levels, which suppress fetal parathyroid function. Being suddenly deprived of maternal calcium, the newborn can become hypocalcaemic, leading to tetany and even death. Definitive treatment is exploratory neck surgery, ideally in the second trimester.

Hypoparathyroidism

The major cause of hypoparathyroidism in pregnancy is the inadvertent removal of the parathyroid gland at thyroid surgery. If untreated, chronic maternal hypocalcaemia causes fetal parathyroid overactivity, bone demineralisation and growth restriction, which may be fatal. Treatment is with vitamin D and calcium, increasing the dosage as pregnancy advances, with close monitoring of calcium and phosphate levels. Hypercalcaemia from overdosage is a risk with vitamin D analogues such as calcitriol. Infants receiving breast milk from mothers consuming large doses of vitamin D should have periodic measurements of calcium levels.

Inherited Metabolic Diseases

Phenylketonuria is an autosomally inherited absence of phenylalanine hydroxylase. Severe mental retardation occurs in untreated children. Treatment is with a diet low in phenylalanine. This diet should be reintroduced before conception to reduce the risk of miscarriage, growth retardation, congenital heart disease, microcephaly and mental retardation in the offspring. Newborns are screened by means of the Guthrie test, which detects excess phenylalanine in the blood. Women with treated phenylketonuria can be allowed to breast feed.

Marfan's syndrome is characterised by aortic aneurysm, joint laxity and kyphoscoliosis. Musculoskeletal effects rarely complicate pregnancy, but there is an increased risk of spontaneous miscarriage and low birthweight; aortic dissection and spontaneous uterine inversion may occur.

Women with osteogenesis imperfecta may have blue sclerae, otosclerosis and fragile bones that fracture easily. Skeletal deformities cause cephalopelvic disproportion and often necessitate caesarean section. Calcium supplements during pregnancy are advisable.

Achondroplasia (dwarfism) can cause pelvic contracture that may necessitate caesarean delivery. Cervical spinal stenosis can lead to difficulty with tracheal intubation; epidural may also be challenging.

Gastrointestinal Disease

Gastrointestinal Tract Changes in Pregnancy

Appetite and calorie intake increase in pregnancy. Dietary aversions and pica (mouthing of non-food items) may occur. Up to 80% of uncomplicated pregnancies are associated with nausea and vomiting, maximal between the 6th and 14th weeks; in 20% of cases symptoms persist throughout pregnancy. Attention to fluid balance, small meals and avoidance of potential irritants are recommended. Judicious use of metoclopramide, meclizine or prochlorperazine is safe (FDA Category B: no evidence of risk in humans).

Oedema and hyperplasia of the gums with excessive contact bleeding are almost universal. The incidence of caries is increased twofold in pregnancy. Gastric acid and pepsin secretion are reduced, as are muscle tone and motility, while water reabsorption is increased, predisposing to constipation. Some of the clinical signs of liver disease such as palmar erythema and spider naevi are normal in pregnancy, but jaundice, liver enlargement and tenderness are not. Total protein and albumin decrease during pregnancy, while globulins, fibrinogen, cholesterol, triglycerides, alkaline phosphatase, bile acids and factors VI, X and XI are increased; total bilirubin, transaminases, γ-glutamyl transferase, prothrombin and partial thromboplastin time remain unchanged.

Heartburn

Almost 50% of pregnant women have heartburn in the third trimester. Standard antireflux measures such as small meals, careful diet, attention to posture, smoking cessation and antacids may help. Cimetidine and ranitidine appear safe during pregnancy (FDA category B: no evidence of risk in humans).

Constipation

Apart from reduced gut peristalsis of pregnancy, important causes of constipation include inadequate diet and iron supplements. Treatment includes increasing fluids and fibre. If necessary, methylcellulose, magnesium hydroxide or senna preparations may safely be taken in pregnancy. Liquid paraffin, which impairs absorption of fat-soluble vitamins and phenolphthalein, which has a laxative effect in infants, are contraindicated in pregnant or lactating women.

Haemorrhoids

Increased intra-abdominal pressure and hormonal factors aggravated by multiparity and constipation are responsible for the increased incidence of symptomatic haemorrhoids in pregnancy especially in the puerperium. About 95% have external haemorrhoids, 35% also have internal haemorrhoids, 20% have prolapsed or thrombosed haemorrhoids but only 5% have fissure-in-ano.

Most haemorrhoids present with pruritus, pain and bleeding. Differential diagnosis of pruritus ani includes fungal or parasitic disease, recent use of antibiotics or other topical irritants and diabetes. Management includes avoiding constipation, astringent suppositories (e.g. Anusol) or topical anaesthetics with (e.g. proctofoam) or without (e.g. dibucaine) 1% hydrocortisone. Anal dilatation or emergency haemorrhoidectomy may be required in severe cases. Haemorrhoids usually improve post partum.

Infective Diarrhoea

Most bouts of acute diarrhoea are self-limited. Recent travel to areas of endemic diarrhoeal illness, fever and bloody stools lasting more than 3 days are indications for stool culture and microscopy for ova, cysts and parasites. Treatment does not generally differ from that of non-pregnant patients except that some antibiotics, e.g. tetracyclines, are contraindicated in pregnancy.

Peptic Ulcer Disease

Peptic ulceration rarely arises de novo in pregnancy. Pre-existing disease improves in the majority of pregnant patients. Most complications occur in the puerperium. Aluminium, calcium and magnesium antacids and H2 receptor antagonists are safe. Pregnancy outcome is generally good except when haemorrhage or perforation occur.

Coeliac Disease

Subtotal villous atrophy of the small intestinal mucosa due to dietary gluten occurs in 0.03% of the general population. Untreated patients (both male and female) may be infertile. Pregnancy (and other stressors) may precipitate symptoms in patients with pre-existing disease. Miscarriage, megaloblastic anaemia and intrauterine growth retardation are the major problems in pregnancy. Management includes continuing the gluten-free diet in pregnancy as well as iron, folic acid and zinc supplements.

Inflammatory Bowel Disease

Fertility is unimpaired in most women with well-controlled ulcerative colitis or Crohn's disease, but sulphasalazine is a reversible cause of male infertility. Pre-existing disease may relapse in pregnancy. Fibreoptic endoscopy is safe in pregnancy. Folic acid, vitamin B12, iron, zinc and vitamin D supplements are recommended. Delivery by caesarean section is for obstetric indications only. Breast feeding is not contraindicated.

Disease activity determines pregnancy outcome. Spontaneous miscarriage or stillbirth occurs in up to two thirds of patients with severe active inflammatory bowel disease or those requiring surgery during pregnancy. Maternal mortality is also higher in this group. There is no increased risk of congenital abnormalities. The outcome of one pregnancy does not seem to influence that of later pregnancies.

Jaundice in Pregnancy

Jaundice occurs once in every 1500 pregnancies. The causes may be unrelated to pregnancy, e.g. viral hepatitis, hepatotoxic drugs (e.g. chlorpromazine, tetracycline, steroids), chronic liver disease, gallstones, chronic haemolysis or late HIV infection and acquired immune deficiency syndrome (AIDS). Causes peculiar to pregnancy include intrahepatic cholestasis, hyperemesis gravidarum, pre-eclampsia or eclampsia, the HELLP syndrome (haemolytic anaemia, elevated liver transaminases, low platelet count), acute fatty liver of pregnancy and hepatic rupture (Table 16.7).

Endocrine Diseases

Maternal and Fetal Glucose Homeostasis

Insulin secretion after a glucose load doubles from the first to the third trimesters (insulin resistance). Glucose but not insulin crosses the placenta by facilitated diffusion. Fetal glucose stimulates insulin and inhibits glucagon secretion. Amino acids also stimulate fetal insulin secretion. Maternal ketones and free fatty acids also cross the placenta. Fetal hyperglycaemia stimulates pancreatic beta cell hypertrophy, resulting in hyperinsulinism in utero.

Overt Diabetes (Insulin and Non-insulin Dependent)

Prepregnancy counselling followed by team care ensures tight control of diabetes and reduces the rate of congenital abnormalities, which is approximately double that in non-diabetics. Glycosylated haemoglobin A1 reflects long-term glycaemic control. The higher the haemoglobin A1 levels in the first trimester, the greater the risk of severe congenital abnormalities (neural tube defects, transposition of great vessels, ventricular septal defects, caudal regression syndrome, exomphalos, pulmonary hypoplasia, Potters's syndrome). Ultrasound dating and exclusion of major anomalies begins in the first trimester. This is followed by fetal anomaly

Table 14.7. Clinical features of liver disease in pregnancy. Note that liver biopsy is safe in pregnancy, provided blood clotting is normal; it is rarely indicated

Condition	Effect of/on pregnancy	Investigations and treatment in pregnancy
Intrahepatic cholestasis	Pruritus and jaundice in T2/T3, resolves postnatally and usually recurs. Increased incidence of preterm labour, PPH, perinatal mortality and morbidity	Mild elevation in BR, transaminases, alkaline phosphatase; marked increase in bile acids. Cholestyramine for pruritus; vitamin K to mother and infant. Avoid OCPs
Severe pre-eclampsia and eclampsia	Many liver diseases occurring in late pregnancy are associated with hypertension. 15% have HELLP syndrome	Elevated unconjugated BR and creatinine, microangiopathic haemolytic anaemia with DIC. Transfer to tertiary centre with intensive care facilities
Liver rupture	RUQ pain and shock in older multipara at term precipitated by PET, eclampsia, and HELLP syndrome or trauma; uterine contractions, vomiting, convulsions; 60% maternal and fetal mortality	US, paracentesis, CT and MRI; surgery; prevention and control of hepatic/renal failure and sepsis
Acute fatty liver (rare, aetiology unknown)	Abdominal pain, jaundice, vomiting, headache; may progress to hepatic and renal failure, coagulopathy, coma; 25% maternal and fetal mortality; recurrence is unusual	Neutrophil leucocytosis, hypoglycaemia, metabolic acidosis, raised urea, creatinine, uric acid; moderate elevation of BR and all liver enzymes, low platelets; early delivery by caesarean section
Acute viral hepatitis (A to E, EBV, CMV and herpes simplex). Commonest cause of jaundice in pregnancy	Incidence/severity unaffected by pregnancy; no increase in miscarriage, IUGR, stillbirth or congenital anomalies; high risk of vertical transmission of e Ag positive HB from mother to child. 90% of those born of HBeAg seropositive DNA positive carriers will become chronic carriers	Marked elevation of transaminases; alkaline phosphatase moderately elevated; abnormal clotting suggests fulminant hepatitis; infants of HBsAg positive mothers are given HB immune globulin at birth. In USA, universal prenatal screening for HBsAg and immunisation of all babies beginning at birth
Chronic hepatitis		
A. Chronic active	Amenorrhoea and infertility if untreated; liver function usually preserved during pregnancy; UTIs, prematurity, LBW, PET and fetal loss more common	Anti-smooth muscle Ab positive; continue steroids and/or azothioprine in pregnancy
B. Chronic persistent	No increase in maternal/fetal morbidity, but may progress to CAH or cirrhosis	
Cirrhosis	Perinatal mortality and PPH more common; maternal prognosis depends on hepatic dysfunction; oesophageal varices more likely to bleed	Consider TOP if advanced and decompensated; high carbohydrate, low protein diet and vitamins; shorten 2nd stage; encephalopathy precipitated by infection, excess blood loss, opiates
Primary biliary cirrhosis	Liver function, maternal and fetal outcomes are variable	Anti-mitochondrial Ab positive; penicillamine, cholestyramine and vitamin K

Table 14.7. (continued)

Condition	Effect of/on pregnancy	Investigations and treatment in pregnancy
Familial non-haemolytic jaundice	Pregnancy may increase jaundice; good maternal/fetal prognosis in Gilbert's disease	High levels of unconjugated (Gilbert's disease) and conjugated (Dubin–Johnson and Rotor syndrome) BR; no treatment required
Cholelithiasis	Acute cholecystitis requiring surgery in 1 in 1000 deliveries; concurrent pancreatitis increases maternal/fetal mortality	US or Technetium 99 scan; fetal outcome optimal following cholecystectomy in T2

BR, bilirubin; T2, second trimester; T3, third trimester; PPH, postpartum haemorrhage; OCP, oral contraceptive pill; HELLP, haemolytic anaemia, elevated liver transaminases, low platelet count; DIC, disseminated intravascular coagulation; RUQ, right upper quadrant; PET, pre-eclamptic toxaemia; CT, computerised tomography; MRI, magnetic resonance imaging; EBV, Epstein–Barr virus; CMV, cytomegalovirus; IUGR, intrauterine growth retardation; UTI, urinary tract infection; HB, hepatitis B; LBW, low birthweight; CAH, chronic active hepatitis; Ab, antibody; Ag, antigen.

scanning and maternal serum AFP screening in the second trimester, and fetal surveillance (non-stress test, biophysical profile and umbilical arterial Doppler velicometry) in the third trimester.

Established diabetics on oral hypoglycaemics are changed to human insulin in early pregnancy which is less antigenic and improves metabolic control. Twice daily short and medium-acting insulin is replaced in the second half of pregnancy by preprandial soluble insulin with a long-acting insulin in the evening. Insulin requirements may increase during pregnancy but decrease in the last few weeks, causing nocturnal hypoglycaemia. Continuous subcutaneous insulin pumps are safe and effective in pregnancy. Preprandial blood glucose monitoring at home reduces the necessity for hospital admission.

A quarter of established diabetics have background retinopathy, the incidence rising to 90% in women with nephropathy. (Table 14.8). Rapid induction of glycaemic control in early pregnancy stimulates retinal vascular proliferation. The duration of diabetes and the condition of the retina at the beginning of the pregnancy influences the rate of acceleration. Laser treatment is effective for proliferative retinopathy in pregnancy. Nephropathy occurs in 3–4% of pregnant diabetics and is defined as albumin excretion greater than 300 mg/day or total protein excretion greater than 500 mg/day. These patients are at increased risk of anaemia, pre-eclampsia, growth retardation, fetal distress, preterm delivery and perinatal death. Dialysis may be required to maintain electrolyte balance and prevent metabolic acidosis.

Hypertensive disorders occur in 15–30% of pregnant diabetics, but coronary heart disease is rare, albeit associated with very high maternal mortality and perinatal loss. Asymmetric growth retardation occurs more commonly in diabetics with vasculopathy. Insulin-dependent diabetics are at increased risk of ketoacidosis during pregnancy. Spontaneous miscarriage, congenital anomalies, and polyhydramnios are more common in poorly controlled diabetics. Macrosomia, defined as birthweight above the 90th percentile for gestational age or greater than 4.5 kg occurs in almost one third of diabetic pregnancies. Predictive formulas based on ultrasonic measurements of size all have significant false-positive

Table 14.8. Classification of diabetes in pregnancy. White classification is not widely used in the UK and has been replaced by a simplified version

Group	Simplified classification	Group	White classification
1	Impaired glucose tolerance	A	Asymptomatic diabetes shown by GTT; diet control; any duration or onset
2	Insulin-dependent diabetes < 10 years' duration	B	Onset after age 20, duration < 10 years, no vascular complications
3	Insulin-dependent diabetes, 10-19 years' duration	C	Onset and duration between ages 10-19, no vascular complications
4	Insulin-dependent diabetes > 20 years' duration	D	Onset before age 10, duration 20+ years and background retinopathy or hypertension
		F	Diabetes with nephropathy
		R	Diabetes with retinopathy
		H	Clinically evident atherosclerotic heart disease
		T	Prior renal transplantation

GTT, glucose tolerance test.

(30-50%) and predictive value positive (60-85%) rates. Macrosomia predisposes to traumatic vaginal delivery and shoulder impaction.

Diabetics have a higher risk of preterm labour. Both beta sympathomimetics and corticosteroids cause hyperglycaemia; they may be used under insulin cover, though magnesium sulphate is better tolerated as a tocolytic agent. The perinatal mortality rate among diabetic women remains approximately twice that in non-diabetics. Congenital anomalies are the most common cause of death (approximately 5%). Neonatal morbidity (respiratory distress, hypoglycaemia, hypocalcaemia, polycythemia, jaundice or cardiomyopathy) occurs in 30-50% of newborns. If a parent has type I diabetes, the risk of diabetes in the offspring is 1-6%. The risk for the offspring of a type II diabetic is for type II diabetes only.

Management of Labour

Vaginal delivery at term is attempted if there is a good obstetric history, the ultrasonically estimated fetal weight is less than 3.5 kg, the head is engaged and diabetic control is good. The remaining cases are usually delivered by elective caesarean section. Glucose (5-10 g/h) and insulin (1-2 units/h) are administered intravenously to maintain glycaemic control at 6-7 mmol/l; the infusion is stopped after delivery. Breast feeding is not contraindicated except in those taking oral hypoglycaemic agents.

Gestational Diabetes

This is defined as abnormal glucose tolerance occurring for the first time during pregnancy. The incidence is 10 times higher than that of overt diabetes. The overall prevalence is about 3%, but it increases with age and is higher among blacks, Hispanics and Asians. Since only a quarter of pregnant women have recognisable risk factors, most authorities recommend a full glucose tolerance test in all women with a glucose level greater than 7.8 mmol/l taken 1 h after a 50 g oral glucose load at 28-34 weeks (the mini glucose tolerance test). A positive screening test without a confirmatory abnormal glucose tolerance test does not establish the diagnosis of gestational diabetes. Diagnostic criteria for gestational diabetes vary widely (Table 14.9).

Table 14.9. Diagnostic criteria for gestational diabetes

	O'Sullivan and Mahan	O'Sullivan	British Diabetic Association	WHO[a] Gestational DM	US NDDG Gestational IGT	US NDDG Gestational DM	Id GT	Diabetes
Glucose dose (g)	100	50	50	75	75	100	75	75
Fasting BS (mmol/l)	5.8		6.7	>7.78 (8)	<7.78 (8)	9.17	<7.78	>7.78
1 h BS	10.6	>8.3	10			10.56	>11.11	>11.11
2 h BS	9.2		6.7	>11.11 (11)	7.78–11.06 (8–11)	9.17	7.78–11.06	>11.11
3 h BS	8.1					8.06		
Criteria	Two or more values must be met		Fasting or 2 h value >6.7 and 1 h >10 must be met	Either fasting or 1 h value must be met	Both values must be met	Two or more values must be met	All values must be met	Either fasting or 1 h and 2 h values must be >11.11

[a] The WHO recommendation does not distinguish between pregnant and non-pregnant women.
US NDDG, US National Diabetes Data Group; BS, blood sugar; DM, diabetes mellitus; IGT, impaired glucose tolerance.

Treatment is with insulin when hyperglycaemia is not controlled by a modified carbohydrate diet. Though maternal morbidity is less frequent and severe than in overt diabetes, there is no less risk of macrosomia, birth trauma or neonatal morbidity. There is no increased risk of congenital malformations, but they have a higher than normal perinatal mortality rate.

The glucose tolerance test is repeated 6 weeks after delivery to determine whether she has persistent diabetes. Approximately 50% of gestational diabetics will develop non-insulin-dependent diabetes (type II) in later life, the incidence rising to 70% in obese women. The incidence of juvenile diabetes in their offspring is increased 20-fold over that in the general population.

Thyroid Disease

Maternal and Fetal Thyroid Function in Pregnancy

Increased thyroxine synthesis during pregnancy makes greater demands on iodine reserves (Figure 14.1). Thyroid gland hypertrophy (goitre) is particularly common in areas with dietary iodine deficiency. Adding iodised salt to the diet can prevent iodine deficiency. Severe uncorrected iodine deficiency in pregnancy may result in cretinism (mental retardation, lethargy, large tongue, hoarse cry, coarse skin) and increased rates of congenital anomalies and perinatal mortality. Neonatal screening programmes diagnose most cases of congenital hypothyroidism.

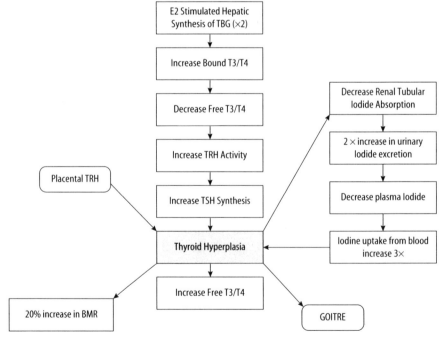

Fig. 14.1. Physiological changes in thyroid function in pregnancy. PGE2, prostaglandin E2; TRH, thyroid-stimulating hormone; TBG, thyroid-binding globulin; TSH, thyroid-stimulating hormone; BMR, basal metabolic rate.

The fetal hypothalamo-pituitary-thyroid system develops and functions autonomously. Fetal thyroxine is synthesised by the end of the first trimester, the thyroid becoming vulnerable to placental passage of goitrogens. Both thyroid-stimulating hormone (TSH) and thyroxine levels increase between 18 and 22 weeks, after which TSH levels off, while thyroxine continues to rise until term. Because of a TSH surge, physiological hyperthyroidism occurs after birth, returning to normal within a week.

Hypothyroidism

Hypothyroidism leads to anovulation, lethargy, and cold intolerance; symptoms are masked by the hypermetabolism of pregnancy. It occurs in 0.6% of pregnant women. Common causes in pregnancy include autoimmune thyroiditis, and prior treatment with radioactive iodine or thyroidectomy. There is an increased risk of spontaneous miscarriage, congenital anomalies and perinatal mortality. Diagnosis is hampered by oestrogen-induced doubling of thyroid-binding globulin, which leads to apparently adequate thyroxine levels in the presence of low free thyroxine. The only definitive biochemical tests are measurement of free thyroxine and TSH (Table 14.10). Severe hypothyroidism is treated with replacement doses of thyroxine (2 μg/kg), monitoring free thyroxine and TSH levels to adjust the dose. Breast feeding is not contraindicated.

Hyperthyroidism

Hyperthyroidism occurs in 0.2% of pregnant women. The clinical features (warm extremities, heat intolerance, tachycardia, systolic flow murmur) are similar to those of normal pregnancy. Most women with thyrotoxicosis in pregnancy have Grave's disease. Thyrotoxicosis can also present as hyperemesis gravidarum. Untreated hyperthyroidism is associated with maternal heart failure, congenital anomalies and significantly higher perinatal mortality. Diagnosis is hampered by physiological alterations in biochemical tests (Table 16.10). Elevated free thyroxine (T_4) and low TSH levels are diagnostic. Free T_3 levels are measured if the diagnosis remains unclear.

The disease is best treated prior to conception; carbimazole or propylthiouracil inhibit thyroid hormone synthesis. Tapering the dose as pregnancy progresses minimises the risk of neonatal goitre and hypothyroidism resulting from transplacental passage of the drugs. Agranulocytosis is rare, but reversible on stopping the drug. Close observation is essential to avoid a thyroid crisis in the puerperium.

Table 14.10. Changes in thyroid function tests in pregnancy

Test	Normal pregnancy	Hypothyroid	Hyperthyroid
T_3RU	< 10%	Increased	Decreased
RIA Total T_4 (mmol/l)	90–200	Decreased	Increased
RIA Total T_3 (ng/dl)	50% increase	Decreased	increased
Free T_4 (pmol/l)	Unchanged		
Free T_3 (pmol/l)	Unchanged		
$FT_4I = T_4 \times T_3RU$	Unchanged	Decreased	Increased

RIA, radioimmunoassay; T_3, T_4, Thyroxin; RU, resin uptake; FT_4I, free thyroxine uptake.

This is treated with beta blockers, iodides and carbimazole. Apart from iodides, antithyroid drugs are safe, in low doses, for breast feeding mothers. Thyroidectomy is rarely necessary in pregnancy except for severe pressure symptoms; radioactive iodine is contraindicated altogether.

In about 1% of women with Grave's disease or Hashimoto's thyroiditis, thyroid-stimulating immunoglobulins cross the placenta, activate the thyroid gland and result in fetal or neonatal thyroid dysfunction. Cordocentesis may be used to confirm the diagnosis and monitor treatment.

Postpartum Thyroiditis

Mild transient hyperthyroidism about 3 months after delivery, followed by hypothyroidism and often a goitre, occurs in 5–10% of individuals. It may be a rebound phenomenon, the new mother producing thyroid antibodies after the immunosuppressive effect of pregnancy is removed. Treatment is symptomatic; spontaneous remission often occurs. There is a high risk of recurrence after subsequent pregnancies; permanent hypothyroidism and other autoimmune diseases may also develop.

Adrenal Disease

Cortisol and aldosterone increase in pregnancy under the influence of oestrogen-induced stimulation in binding globulin and renin secretion, respectively.

Cushing's Syndrome

Adrenocortical hyperfunction causes infertility; it is exceedingly rare in pregnancy. Clinical features include acne, striae, weight gain, oedema and weakness, all of which may occur in normal pregnancy. Diagnosis is hampered by pregnancy-induced changes in free and bound cortisol levels. The diagnosis is excluded if the morning plasma cortisol can be suppressed by dexamethasone. Treatment depends on the primary cause: microsurgery for a pituitary adenoma or exploratory surgery if an adrenal tumour is suspected. Metyrapone blocks cortisol secretion to reduce hypercortisolism until fetal maturity is attained. Miscarriage, diabetes and hypertension with consequent growth retardation, premature labour and intrauterine death are common. Adrenal failure is a risk in the newborn.

Addison's Disease

Untreated patients are often amenorrhoeic and infertile. Clinical features (tiredness, nausea, vomiting, hyperpigmentation) overlap those of pregnancy. The diagnosis is confirmed by a low plasma cortisol level, which does not rise after synthetic adrenocorticotrophic hormone (ACTH). Vomiting, infections and the stress of labour carry serious maternal risk in the untreated patient. Treatment with replacement doses of hydrocortisone and 9-α-fluorohydrocortisone reduces maternal and fetal risk, of which growth retardation is the most significant. Delivery requires supplemental doses of hydrocortisone.

Phaeochromocytoma

Phaeochromocytoma is a rare adrenal medullary tumour (90% are unilateral) that may present with intermittent, labile hypertension, or paroxysmal symptoms such as anxiety, headaches and palpitations. The diagnosis is confirmed by elevated levels of 24 h urinary vanillymandelic acid or metanephrines and plasma catecholamines. Hypertension is controlled with alpha blockade (e.g. phentolamine) followed by beta blockade (e.g. propranolol). Surgical removal after localisation with magnetic resonance imaging may be required during pregnancy. Maternal (0–17%) and fetal mortality (15–35%) have improved.

Pituitary disease

Pituitary tumours

Prolactin-secreting adenomas are the most common pituitary tumours encountered in pregnancy. Unlike women with untreated macroadenomas, most patients with microadenomas have uncomplicated pregnancies. There is no increase in the rate of miscarriage, congenital abnormality or multiple pregnancy in symptomatic women taking bromocriptine. Pituitary fossa enlargement can be detected by measurements of prolactin and visual field testing (for bitemporal hemianopia). This complication regresses after delivery. Labour generally starts and proceeds normally. Breast feeding does not cause pituitary tumours to enlarge, but prolactin levels do not rise on suckling.

Hypopituitarism

The anterior pituitary doubles in size during pregnancy. Avascular necrosis of the anterior pituitary leading to hypopituitarism may be caused by severe postpartum haemorrhage (Sheehan's syndrome). Insulin-dependent diabetics may also develop antepartum pituitary infarction. Pregnancy outcome after treated hypopituitarism is often successful. In untreated cases, miscarriage, stillbirth and maternal death may result.

Diabetes Insipidus

The incidence of diabetes insipidus in pregnancy (1 in 15 000) is about the same as in the general population. It may be idiopathic or caused by tumour expansion or skull trauma. Vascular anastomoses protect the posterior pituitary from avascular necrosis. Pregnancy may worsen the condition. Treatment is with synthetic vasopressin analogues.

Acute Abdominal Pain in Pregnancy

Obstetric (e.g. placental abruption) and gynaecological causes (e.g. early pregnancy complications, fibroid degeneration) of abdominal pain in pregnancy are not considered here. Pressure symptoms (fetal movements, head engagement, abdominal wall distension) and Braxton Hicks contractions (irregular in intensity and frequency) are common causes of pain in the third trimester. The most important renal cause of abdominal pain is pyelonephritis. Vascular accidents (e.g. in sickle cell crisis), malignant disease and porphyria are rare causes of abdominal pain in pregnancy. The following conditions may occur at any time during pregnancy.

Acute Appendicitis

Acute appendicitis is the most common non-obstetric surgical emergency occurring in 0.01% of all pregnancies. Differential diagnosis from pyelonephritis, cholecystitis, ruptured ovarian cyst, degenerating fibroid, ectopic pregnancy, placental abruption, mesenteric adenitis, inflammatory bowel disease, intestinal obstruction and diverticulitis may be difficult. The position of the appendix will determine the location of pain and tenderness on examination. Rebound tenderness is absent in one third of patients in the third trimester.

Laparoscopy (in the first trimester) and/or laparotomy are indicated regardless of the gestational age. A midline, paramedian or muscle-splitting incision over the point of maximum tenderness is generally adequate in the second half of pregnancy. Simultaneous caesarean delivery for obstetric reasons has no additional morbidity. Broad spectrum intravenous antibiotics and peritoneal drainage are required for perforation, peritonitis or abscess formation, which are associated with higher perinatal and maternal mortality. Uncomplicated appendicitis is associated with a 5% risk of spontaneous miscarriage or preterm labour.

Ureteric Obstruction

Renal and ureteric calculi are the major causes of ureteric colic, occurring in 0.03% of pregnancies. Ultrasound may demonstrate unilateral dilatation of the collecting system. If the diagnosis is still in doubt a one shot pyelogram is permissible in pregnancy. Analgesia and a high fluid intake may help, as 50% of stones pass spontaneously. Any infection is treated with the appropriate antibiotics. Surgical intervention is required in 5% of cases (Table 14.6).

Acute Pancreatitis

Acute pancreatitis occurs in 0.01–0.001% of births. Symptoms and signs (nausea, vomiting, epigastric pain radiating to the back, fever and shock), diagnostic tests (elevated amylase and glucose, polymorphonuclear leucocytosis, and hypocalcaemia) and treatment (analgesia, nasogastric suction, and correction of fluid and biochemical disturbances) are similar to that in non-pregnant patients. Uncomplicated pancreatitis is generally self-limiting but it may recur. Maternal mortality is no greater than in non-pregnant women, but fetal losses are about 10%. Termination of pregnancy has no effect on maternal outcome.

Acute Intestinal Obstruction

Adhesions are the most common cause of intestinal obstruction, followed by volvulus, hernia and tumour. Obstruction is most likely in the second half of pregnancy as the uterus becomes an abdominal organ. Though symptoms such as nausea, vomiting, abdominal pain, distension and constipation occur in normal pregnancy, intestinal obstruction is a possibility when they occur simultaneously. Distended loops of bowel are seen on plain abdominal radiograph. Laparotomy should be carried out if correction of fluid and electrolyte balance and nasogastric decompression are not successful.

Malignant Disease

Malignant disease occurs in 0.01% of pregnancies. With the exception of malignant melanoma, pregnancy does not usually affect the natural history of cancer. Transplacental passage of metastases to the fetus is rare. The patient should be treated as though she were not pregnant. Systemic chemotherapy may have teratogenic effects and radiotherapy almost always leads to fetal death and miscarriage. Carcinoma of the cervix is the most frequently diagnosed invasive neoplasm in pregnancy (about 1 in 2000 pregnancies) and is covered on p. 173. Cancer of the vulva, vagina and endometrium are extremely rare in pregnancy. The clinical features of selected malignant diseases in pregnancy are described in Table 14.11.

Table 14.11. Clinical features of selected malignant disease in pregnancy

Type	Effect on/of pregnancy	Treatment in pregnancy
Rectum/colon	1 in 50 000 pregnancies; ulcerative colitis and familial polyposis coli are predisposing factors; complications if surgery is delayed to allow fetal viability	Radical surgery in T1 and T2; may be deferred until the fetus is viable in T3; delivery usually by C/S
Haematological 1. Hodgkin's lymphoma	1 in 6000 deliveries; pregnancy does not adversely affect outcome	< 20 weeks: treatment controversial but might include TOP. > 20 weeks: most advocate delaying chemo- and radiotherapy until after delivery
2. Leukaemia (a) Acute (90%): usually AML	1 in 75 000 pregnancies; pregnancy has no effect on outcome; increased risk of haemorrhage, infection and abortion secondary to disease and treatment	< 20 weeks: consider TOP and aggressive chemotherapy. > 20 weeks: aggressive chemotherapy
(b) Chronic (10%): usually CML	Early chemotherapy does not influence survival or outcome	Chemotherapy can be withheld until after delivery; long-term follow-up of offspring
Malignant melanoma	Pregnancy accelerates metastatic spread but no significant difference in prognosis (at least for stage I disease); delay future pregnancies by several years; avoid OCPs	Advanced stage disease < 20 weeks: consider TOP and wide excision with regional lymph node dissection and skin graft. > 20 weeks: may defer chemo/radiotherapy until after delivery
Thyroid carcinoma	Avoid thyroid scan because of risk of fetal irradiation; pregnancy does not usually affect these usually slow growing tumours	Total thyroidectomy and lymph node dissection; replacement thyroxine to suppress TSH

T1, T2 and T3, first, second and third trimester; TOP, termination of pregnancy; AML, acute myeloid leukaemia; CML, chronic myeloid leukaemia; OCP, oral contraceptive pill; TSH, thyroid stimulating hormone; c/s, caesarean.

Breast Cancer

Breast cancer is uncommon under the age of 35 years; it occurs in 0.03% of pregnancies. The overall risk is double in women whose mother or sister have had the disease. Pregnant patients tend to present with more advanced disease than non-pregnant women. Mammography is seldom useful during pregnancy. Treatment is either by simple mastectomy and lymph node dissection or lumpectomy with lymph node sampling, followed by chemotherapy if nodes are positive. A short delay prior to treatment in the second trimester while awaiting fetal viability is permissible. Breast feeding is generally discouraged, though there is no convincing evidence that nursing adversely affects prognosis. A 2–3-year interval is recommended between therapy and subsequent pregnancy.

Women with small (less than 2 cm) localised tumours have an 85% 5-year survival, falling to 25% if lymph nodes are positive. Therapeutic abortion does not improve survival.

Ovarian Cancer

Ovarian cancer is exceedingly rare (one in 20 000 to 30 000) in pregnancy. About 90% of functional cysts will regress spontaneously by the second trimester. All remaining ovarian masses should be removed at exploratory laparotomy through a midline incision. Most ovarian malignancies in pregnancy are diagnosed at stage I. Minimal therapy in early pregnancy is ovarian cystectomy or oophorectomy and biopsy of the other ovary with peritoneal washings. If disease is advanced (one third are stage III or IV at diagnosis), bulk-reducing surgery followed by aggressive chemotherapy are essential. In the third trimester exploratory surgery may be combined with caesarean section if the fetus is viable.

There is a variable risk of miscarriage, torsion (10%), rupture (8%), intracystic haemorrhage (5%) and obstructed labour. The overall 5-year survival rate in ovarian cancer is 75%; as in non-pregnant women it is related to the stage and histological type of tumour. Pregnancy does not adversely affect the prognosis.

Skin Disease

Physiological Changes During Normal Pregnancy

Chloasma, linea nigra, and other localised areas of hyperpigmentation (areola, axilla, perineum and pre-existing naevi) are common in pregnancy, even in those with a fair complexion. Vascular changes, e.g. spider angiomas (60%), haemangiomas (5%) and palmar erythema also occur. Striae develop in 90% of pregnant women; they tend to be familial. Nails become soft and brittle. Increased sweat and sebum excretion occur in the third trimester. More hair follicles enter the active phase (anagen) of growth during pregnancy, creating thicker hair. The resting phase (telogen) starts soon after delivery, causing hair to fall for several months (telogen effluvium). Reassurance is all important as hair density returns to normal spontaneously. Skin conditions specific to pregnancy include the following.

Pruritus Gravidarum

Intense pruritus, which may occur anywhere, but is usually more severe on the extremities, occurs in about 2% of all pregnancies, usually in the third trimester, remitting after delivery. Most cases are related to intrahepatic cholestasis, when preterm labour and low birthweight are more common. Other causes include infestations (scabies, lice), acute hepatitis, drug reactions and chronic diseases. Treatment is with emollients and antihistamines if required. Cholestyramine binds bile salts and may help symptomatically. It recurs in about 50% of subsequent pregnancies.

Prurigo Gestationis

Excoriated papules resulting from scratching occur on the abdomen and extensor surfaces of limbs in up to 2% of pregnancies, usually between 25 and 30 weeks' gestation, most commonly in atopic women. The aetiology is unknown, but there is no apparent risk to the mother or fetus. Symptomatic treatment is with topical corticosteroids and antihistamines. Recurrence in subsequent pregnancies is rare.

Pruritic Urticarial Papules and Plaques of Pregnancy

This harmless disorder occurs in about 0.5% of pregnancies, usually in the third trimester. Seventy-five per cent of cases occur in primigravidae; recurrence is rare. Erythematous papules and urticaria-like swollen plaques occur on the abdomen and limbs, but not the face. Topical corticosteroids and antihistamines may help, but the disease promptly clears after delivery.

Papular Dermatitis

Widespread papular lesions may occur over the trunk and extremeties in about 0.03% of pregnancies. The condition may begin at any time during pregnancy, and may be associated with higher fetal mortality. High dose systemic corticosteroids are generally given though most cases improve postpartum. Recurrence in subsequent pregnancies often occurs.

Herpes Gestationalis

This blistering disease (unrelated to the herpes virus) occurs in about 0.01% of pregnancies. It appears in the second and third trimester of pregnancy, occasionally improving before delivery. It presents with intense pruritus and papules (urticaria like), which become vesicles and bullae within a few days, spreading over the whole skin surface. Crusting eventually occurs and healing is usually without scars. Treatment is with systemic steroids. Infants may be affected transiently and they may be preterm or small for dates. Postpartum flare-ups are very common. Recurrence with oral contraceptive use or during subsequent pregnancies, usually more severe and earlier in onset, is common.

Impetigo Herpetiformis

Painful red plaques studded with sterile pustules accompanied by painful oral lesions and systemic effects (fever, arthralgia, splenomegaly and leucocytosis) is very rare but may progress rapidly in pregnancy. It is histopathologically identical with generalised pustular psoriasis. The high risk of maternal death demands treatment with systemic steroids and antibiotics for secondary infection. The disease may recur after pregnancy.

Chronic Skin Disorders in Pregnancy

The effect of pregnancy on selected chronic skin disorders is summarised in Table 14.12. Skin diseases that are usually aggravated by pregnancy include condylomata acuminata, erythema multiforme, erythema nodosum, herpes simplex, lupus erythematosus, pemphigus, pityriasis rosea, porphyria and scleroderma.

Table 14.12. Clinical features of chronic skin disorders in pregnancy

Type	Effect of/on pregnancy	Treatment in pregnancy
Psoriasis	Improves in 50% of pregnant women; may recur post partum	Topical steroids or UV light but not PUVA or methotrexate
Atopic dermatitis	Unpredictable response to pregnancy	Avoid skin irritants; topical emollients, steroids, and antihistamines; antibiotics for secondary infection
Pemphigus vulgaris	May be aggravated by pregnancy; clinically indistinguishable from herpes gestationalis	Systemic steroids
Erythema nodosum	Can be precipitated by pregnancy; pregnancy is unaffected in the absence of precipitating causes with normal chest radiograph and blood count	Nodules generally remit without treatment; extensive painful disease may require steroids
Neurofibromatosis	Progresses rapidly in pregnancy which may also precipitate arterial hypertension	Genetic counselling (autosomal dominant)
Dermatitis herpetiformis	Pregnancy effect is variable	Dapsone may be continued in pregnancy though may cause fetal methaemoglobinaemia
Condylomata acuminata	Can grow rapidly causing introital obstruction; newborn can become infected	Podophyllin is contraindicated in pregnancy; laser or cyrosurgical treatment; delivery by caesarean section for extensive disease at delivery
Acne vulgaris	Pregnancy has a variable effect	Topical benzoyl peroxide, salicylic acid or erythromycin/clindamycin. Tetracycline, sulphonomides and isotretinoin are contraindicated

UV, ultraviolet; PUVA, psoralen and ultraviolet A.

Table 14.13. Alteration in normal haematological values during pregnancy (note that coagulation factors II, VII, IX, X, XII and fibrinogen also increase during pregnancy)

	Prepregnancy	Pregnancy value
Red cell mass (ml)	1400	1650–1850 (20–30% increase)
Plasma volume (ml)	2600	3800–4200 (45–60% increase)
Haemoglobin (g/dl)	12–14	10–12
Haematocrit (%)	40	35
Mean cell volume (fl)	83–85	85–88
Mean cell haemoglobin (pg)	28–29	29
Mean cell haemoglobin concentration (g/dl)	33	34
White blood cells	5.5–6	7–10
Platelets (/μl)	200 000	97 000–150 000
ESR (mm)	< 20	40–100

ESR, erythrocyte sedimentation rate.

Haematological Disorders

Worldwide, haematological disorders account for a large proportion of morbidity during pregnancy. Alterations in normal haematological values are shown in Table 14.13. The physiological fall in haemoglobin and haematocrit in pregnancy is due to a disproportionate rise in plasma volume, compared with red blood cells. The total volume of red cells falls at delivery, reaching non-pregnant values 3 weeks later.

Anaemia

Anaemia in pregnancy, arbitrarily defined by a haemoglobin concentration less than 10 g/dl, may be due to pre-existing causes or it may occur because of (a) deficient red cell production due to poor diet and/or the demands of the pregnancy on iron and folic acid stores, (b) chronic repeated blood loss, and (c) haemolysis of red blood cells. Three quarters of all cases are due to iron deficiency. Most women are asymptomatic and escape detection unless routine screening is performed.

Iron Deficiency Anaemia

Many women begin pregnancy with low iron stores due to previous pregnancies, menstrual loss and poor diet. In developing countries chronic blood loss (e.g. due to chronic hookworm infestation) is a major cause of iron deficiency anaemia. Iron deficiency anaemia occurs in approximately 10% of pregnant women in the UK.

The diagnosis of iron deficiency anaemia is based on a haemoglobin of 10 g/dl or less together with hypochromia, microcytosis and reticulocytosis; serum iron (< 60 μg/dl), ferritin (< 10 ng/dl) and transferrin saturation (< 15%) are low. The total iron binding capacity increases to compensate, but this may be inaccurate because the pregnancy-induced rise in oestrogen and progesterone stimulates transferrin production.

Mild iron deficiency anaemia does not adversely affect pregnancy; infants born to such mothers are seldom iron deficient themselves. However, women who do not receive iron supplements have no stores at the end of pregnancy. If they are also lactating they may remain iron deficient. Severe anaemia increases the risk of fetal death, preterm labour and low birthweight as well as aggravating maternal risk from blood loss at delivery.

Iron deficiency anaemia can be prevented by eating foods with a high iron content. Some prefer routine iron supplements; others give iron only to those who are anaemic. Total iron requirements double in pregnancy from 3 to 6 mg/day. This requires a dietary intake of 60 mg of elemental iron, since only 10% of it is absorbed through the gut. This is equivalent to 300 mg of ferrous sulphate, 200 mg of ferrous fumarate or 100 mg of ferrous succinate per day. Iron is absorbed equally well from any of the salts, though some may be better tolerated than others. Absorption is maximal if supplements are given 30 min before meals. Reticulocytosis should be observed after about a week of therapy. Adequate treatment increases the haemoglobin by 1 g/dl per week. Treatment should continue for another 6 months to replenish body stores. Non-compliance because of unacceptable side-effects (gastric symptoms, constipation or diarrhoea) is the most common cause of treatment failure. The use of parenteral iron (iron dextran or iron sorbitol) is limited by the risk of anaphylaxis. Rarely, blood transfusion may be required for severe anaemia.

Megaloblastic Anaemia

Folic acid deficiency is the major cause of megaloblastic anaemia in pregnancy; vitamin B12 deficiency is very rare except in developing countries or strict vegetarians. Risk factors for folic acid deficiency include multiple or closely spaced pregnancies, haemolytic anaemia, phenytoin therapy and malabsorptive disorders.

Daily folic acid requirements increase threefold during pregnancy and lactation to meet increased maternal and fetal needs. Folic acid deficiency may cause severe symptomatic anaemia (pallor, breathlessness) in women with other signs of malnutrition (skin roughness and glossitis). There is macrocytosis, with leucopenia, hypersegmentation of the white cells and thrombocytopenia. Normal or elevated serum iron and depressed serum folate as well as megaloblasts in a bone marrow aspirate are diagnostic.

Prevention is by encouraging a diet high in folic acid (liver and green vegetables) as well as routine folate supplements (0.3–0.5 mg/day). There is a small risk that, while correcting the anaemia of vitamin B12 deficiency, folic acid therapy may allow subacute combined degeneration of the cord to progress. Proven folic acid deficiency requires a higher oral or intramuscular dose (5–10 mg/day). The response is monitored by a rise in the reticulocyte count after 2–3 days.

Apart from prevention of megaloblastic anaemia there is no evidence that folic acid supplements prevent obstetric complications, but periconceptual folic acid reduces the risk of neural tube defects.

Haemoglobinopathies

The haemoglobinopathies are the commonest inherited (autosomal recessive) diseases worldwide. They include the thalassaemias (alpha and beta), which result from reduced globin synthesis, and haemoglobin variants (sickle cell trait, sickle cell disease and sickle cell/HbC disease) due to structural globin abnormalities.

Each molecule of normal haemoglobin consists of a haem complex surrounded by paired alpha and beta globin chains. Normal adult blood also contains Hb A2 (2%), which has two alpha and two delta chains, and Hb F (< 1%), which has two alpha and two gamma chains. Separate genes control the production of the alpha (four genes), beta, delta and gamma chains (two genes each). Sixty-five per cent of haemoglobin is synthesised in the nucleated red cells and the remainder in the reticulocytes.

1. *Alpha thalassaemia major* is commonest in South East Asia. No alpha chains are made but a tetramer of gamma chains (Hb Barts) is formed in utero. Fetal hydrops due to severe anaemia is usually fatal. Risks to the mother include pre-eclampsia and difficult vaginal delivery due to the enlarged placenta and macrosomic fetus. Deletion of three alpha genes results in moderately severe haemolytic anaemia due to breakdown of haemoglobin H (four beta chains) and Hb Barts (four gamma chains).

   In alpha thalassaemia minor (trait) there is deletion of one (alpha 2) or two (alpha 1) genes. These women do not make any abnormal haemoglobin but develop a hypochromic microcytic anaemia, that must be differentiated from iron deficiency by means of haemoglobin electrophoresis. Genetic counselling and prenatal diagnosis are possible. Regular oral iron and folate supplements are essential during pregnancy but parenteral iron is contraindicated as it may cause iron overload.

2. *Beta thalassaemia* results when defective beta globin genes are inherited. Thalassaemia may be minor (trait) when one gene is abnormal or major (homozygous) when both genes are abnormal. The overall carrier rate in the UK is 1 in 10 000 but it rises to 14 in 100 in the Cypriot population in London. It is also common in the immigrant Asian and black population. All such women should be screened by haemoglobin electrophoresis and a blood film to detect carriers. If their partner also proves to be a carrier, genetic counselling and prenatal diagnosis should be offered as their offspring have a 25% chance of inheriting thalassaemia major.

   Women with thalassaemia minor are often anaemic. The blood film and red cell indices resemble those of iron deficiency but their serum iron and ferritin levels are high. The relative lack of beta chains results in increased Hb A2 and sometimes Hb F. These women require the usual oral iron and folate supplements during pregnancy but parenteral iron is contraindicated as it may cause iron overload. Blood transfusion may be necessary. There is no significant increase in maternal or fetal morbidity.

   Regular transfusion and iron chelation therapy is required to prevent anaemia, congestive cardiac failure, bone deformities and the effects of iron overload, e.g. haemosiderosis, in thalassaemia major. These children usually survive into young adulthood, but pregnancy is rare. If it does occur, folate but never iron supplements should be taken.

   The double heterozygote for Hb S and beta thalassaemia is clinically more severe than either of the traits alone, the severity depending on the proportions of Hb S and Hb A. The sickledex test detects Hb S rapidly without false-negative results but the variants of haemoglobin can be distinguished only by electrophoresis.

3. *Sickle cell anaemia* occurs in 0.3% of the black population of the US, and 10% are heterozygous for the Hb S gene. In the UK (London) the equivalent figure is 9% in black Caribbean women and 17% in those of direct African origin. Women with sickle cell trait are at increased risk of bacteriuria and pyelonephritis during pregnancy, but their maternal and perinatal mortality is unchanged. They are generally asymptomatic but may become iron deficient. Adequate oxygenation during anaesthesia is essential. Cord blood should be checked for Hb S.

In Hb S, glutamine replaces valine at position 6 on the beta chain. The Hb S molecule changes shape when reduced, distorting red cells, which block the microvasculature and may cause painful infarction. Sickling may be precipitated by acidosis, dehydration, hypoxia, fever, infection or severe stress. Meticulous antenatal care, prevention of infection, adequate hydration and regular transfusions for severe anaemia are important. These women are at risk of anaemia, miscarriage, pre-eclampsia, infection, premature labour and painful sickling crises during pregnancy, particularly if there is 50% or more of Hb S present. Oxytocin and epidural anaesthesia are safe, provided anoxia, infection, hypotension and dehydration are avoided.

4. *Homozygous Hb C disease* is rare (1 in 4500 black Americans), and it causes mild haemolytic anaemia with target cells. There is no increased morbidity and mortality associated with pregnancy and no treatment is necessary. Hb C trait is a benign condition causing no obstetric problems. Anaemia is rare. Sickle cell–Hb C disease occurs with the same frequency as sickle cell anaemia in black Americans. Haemoglobin electrophoresis reveals approximately 60% Hb C and 40% Hb S. Painful sickling crises may occur especially in the puerperium. Management is the same as that of sickle cell disease.

Haemolytic Anaemia

The clinical and laboratory features of haemolytic anaemia are summarised in Table 14.14. Spherocytosis is inherited as an autosomal dominant trait. Increased osmotic fragility leads to haemolytic anaemia, which can be precipitated by infection, trauma or pregnancy itself. Folate supplements and vigilance for haemolytic crises are required in pregnant women who have not had a splenectomy. Unless anaemia is severe and untreated, perinatal morbidity and mortality are normal.

The most common inherited (X-linked) enzyme defect is glucose-6-phosphate dehydrogenase deficiency, which occurs in black, mediterranean, and selected Asian and Jewish women. Infection and a variety of oxidant drugs, e.g. aspirin, sulphonamides, nitrofurantoin and antimalarials, may cause red cell haemolysis. The neonate affected with haemolytic anaemia is at risk of hyperbilirubinaemia and kernicterus.

Table 14.14. General features of haemolysis

Clinical	Pallor, jaundice, (mild, fluctuating), spenomegaly, pigment gallstones, leg ulcers
Laboratory	1. Increased red cell breakdown: unconjugated bilirubinaemia, urobilinogen and stercobilinogen increased, haptoglobins absent or reduced
	2. increased red cell production: reticulocytosis and bone marrow erythroid hyperplasia
	3. Damaged red cells: microspherocyte. Increased osmotic fragility, shortened red cell survival

Maternal Diseases in Pregnancy

Autoimmune haemolytic anaemia may be primary or secondary to systemic lupus erythematosus, lymphoid malignancies or drug reactions. The presence of red cell antibodies is detected by the direct Coombs' test. IgM-mediated autoimmune disease poses no risk to the fetus, but most cases are the result of IgG antibodies that cross the placenta and may cause haemolytic anaemia, leucopenia and thrombocytopenia in the fetus. Fetal monitoring is essentially similar to that in rhesus disease. Corticosteroids are indicated for severe symptomatic anaemia in the mother. Splenectomy, immunosuppression or plasmapheresis may be required. It may be difficult to cross-match blood for transfusion. These patients have increased folate requirements in pregnancy (1–3 mg/day).

Aplastic Anaemia

Aplastic anaemia is rare in pregnancy. It presents with pancytopenia. Markedly hypocellular bone marrow is diagnostic. Pregnancy may exacerbate marrow depression and some cases remit spontaneously after termination. Therapeutic abortion does not improve the outcome.

Coagulation Disorders

Table 14.15 lists the results of diagnostic tests in various haemostatic disorders.

Thrombocytopenia

The main causes of thrombocytopenia in pregnancy are DIC and idiopathic thrombocytopenic purpura. Rarely, severe megaloblastic anaemia and marrow depression secondary to drugs, toxic chemicals and malignancy can cause a low platelet count.

Idiopathic thrombocytopenic purpura is an autoimmune disease largely affecting young women of reproductive age, in which IgG antibodies are directed

Table 14.15. Results of laboratory tests in the diagnosis of the major haemostatic disorders (note that in order to avoid testing artefacts the blood should be obtained rapidly and non-traumatically)

Disorder	Platelet count	Bleeding time	Thrombin time	Prothrombin time (extrinsic pathway)	Partial thromboplastin time (intrinsic pathway)
Acquired idiopathic thrombocytopenic purpura	↓	↑	N	N	N
Disseminated intravascular coagulation	↓	↑	↑	↑	↑
Congenital haemophilia A	N	N	N	N	↑
Haemophilia B	N	N	N	N	↑
Von Willebrand's disease	N	↑	N	N	N or ↑

N, normal; ↑, elevated; ↓, decreased.

against the platelets, leading to their sequestration and destruction in the reticuloendothelial system. The diagnosis is based on finding thrombocytopenia with normal red and white cell counts and increased megakaryocytes in the bone marrow. The effect of pregnancy on idiopathic thrombocytopenic purpura is variable; those with active disease before conception are more likely to have haemorrhagic problems than those in remission prior to pregnancy. The fetus is vulnerable to the transplacental effects of the autoantibodies, which may cause fetal and neonatal thrombocytopenia, even in mothers who were in remission during pregnancy. Maternal morbidity and mortality are secondary to haemorrhage during and after delivery. Treatment is with steroids when the platelet count drops below 20×10^9/litre. Alternatives include plasmapheresis, immunosuppressive agents and rarely splenectomy. The risk of intracranial haemorrhage in thrombocytopenic fetuses is not affected by mode of delivery.

Congenital Bleeding Disorders

Haemophilia A (deficiency of factor VIII) is a sex-linked recessive disorder that affects half the male offspring of female carriers. It occurs in 1 in 10 000 of the general population. There is a high mutation rate and in approximately 30% of cases there is no family history. Up to 85% of carriers have a reduced ratio (< 0.7) of factor VIII coagulant activity to factor VIII-related antigen (also known as von Willebrand factor). This is most reliably determined prior to pregnancy, carriers being offered the option of prenatal diagnosis using direct analysis of DNA in fetal blood. Adequate levels of factor VIII can be achieved by giving either cryoprecipitate or factor VIII concentrates. Atraumatic vaginal delivery is preferred.

Factor IX is a vitamin K-dependent clotting factor essential to the intrinsic clotting mechanism. Apart from haemophilia B, low levels of factor IX coagulant clotting factor are also found in liver disease, vitamin K insufficiency (e.g. haemorrhagic disease of the newborn) and during warfarin therapy. Haemophilia B (also known as Christmas disease) is clinically indistinguishable from haemophilia A, though one seventh as common. Prenatal diagnosis is possible. No specific treatment is necessary during pregnancy as factor IX levels tend to rise. Haemorrhage is unusual; it is best treated with fresh frozen plasma or factor IX concentrate.

Von Willebrand's disease is an autosomal dominant inherited coagulation disorder that affects both males and females and consists of a combination of defective platelet adhesion and lack of coagulation factor VIII. Levels of factor VIII coagulant increase normally in pregnancy and determine the likelihood of bleeding, which is greatest in the puerperium. It can be treated with desmopressin, fresh frozen plasma or cryoprecipitate to keep the factor VIII coagulant level at 50% or higher. Von Willebrand's disease does not increase maternal mortality or fetal loss. Prenatal diagnosis is possible but seldom necessary.

Thrombotic Disorders

Pregnancy itself may be considered a secondary thrombotic disorder with increases in fibrinogen and factors VII, VIII and X. Antithrombin III deficiency is a primary thrombotic disorder which occurs in 1 in 2000 of the population. Antithrombin III is the major inhibitor of the serine protease coagulation factors

II, IX, X and XII. Women with hereditary antithrombin III deficiency are at increased risk of blood clotting during pregnancy. They can be managed with infusions of antithrombin III concentrate (or fresh frozen plasma) potentiated by subcutaneous heparin at the time of delivery.

Haematological Malignancies

The incidence and course of these diseases in pregnancy are summarised in Table 14.11.

Chapter 15

Normal Labour

Mechanisms and Course of Labour

Labour is diagnosed when regular painful uterine contractions effect progressive cervical dilatation. Passage of blood-stained mucus (the show) usually occurs early in labour and the membranes rupture late in the first stage. The rate of cervical dilatation is plotted as a sigmoid curve with a latent phase from 0 to 3 cm, followed by an active phase until delivery. In primiparae the cervix dilates at 1–2 cm/h between 1 and 5 cm and 2–3 cm/h between 5 and 10 cm. Progress is judged by dilatation of the cervix and descent of the presenting part (station), i.e. its relationship to the ischial spines, and not on the frequency (normally once every 2–3 min), duration (normally 30–60 s) or intensity of the contractions or whether the membranes have ruptured. Normal labour involves the coordinated action of powers (uterine contractions), passage (bony and soft) and passenger (size, presentation and position of fetus(es)).

Uterine Activity

Myometrial oxytocin receptors (but not maternal oxytocin levels) increase at term. The net effect is stimulation of myometrial activity. Contractions usually arise in the cornua and spread peristaltically from cell to cell; myometrial cells have increased gap junctions facilitating synchronisation.

External tocography (manually or by external pressure transducer) detects changes in the anteroposterior diameter of the abdomen and is used to monitor the frequency and duration but not the intensity of contractions. Internal tocography using a transducer-tipped catheter provides good quality recordings but carries a small risk of infection and fetal injury. Intrauterine pressure measurement is not useful with normal cervimetric progress. Its role is limited to labour in women with previous caesarean section, breech presentation and grand multiparae, especially if oxytocin is to be used.

Labour does not progress adequately unless uterine contractions exceed baseline tone (normally 0.7–1.3 kPa) by at least 2 kPa. The active contraction area

(area under the pressure curve above baseline tone) correlates better than any of the other parameters with the rate of cervical dilatation in the active phase of labour. The rate of rise in pressure is higher in oxytocin-induced contractions than those in spontaneous and prostaglandin-induced labour. Increased intensity and lower frequency contractions occur in the lateral position. Labour in the upright versus recumbent position reduces the use of narcotics and epidural analgesia and shortens the first stage. Ambulation affects progress in spontaneous labour. It decreases the duration of labour, analgesia requirements and incidence of fetal heart abnormalities. The multiparous uterus expends less effort to effect vaginal delivery. In two-third of labours with cephalopelvic disproportion (CPD), the uterine activity is initially greater than in normal labour but later declines.

Pelvis

The normal pelvic shape is gynaecoid (inlet is widest transversely and outlet widest anteroposteriorly (AP)). The anthropoid (flattened oval with widest diameter AP) and android pelvis favour the occipitoposterior position (OPP). Pelvic dimensions are measured at the inlet (AP 11.5 cm, transverse 13.5 cm), midcavity (all diameters 12 cm) and outlet (AP 12.5 cm, transverse 10.5 cm). The thickness and pliability of the soft tissues may alter the functional dimensions of the pelvis.

Fetus

The presenting part is engaged when the widest diameter has passed the plane of the pelvic brim. In most cases, the fetal head presents by the vertex (area between the biparietal eminences and the anterior and posterior fontanelles) and the presenting diameter (suboccipitobregmatic) is 9.5 cm. The deflexed head presents by the occipitofrontal diameter (11.5 cm). As deflexion increases (brow presentation), the mentovertical diameter (13 cm) presents; in face presentation the head is fully extended and the submentobregmatic (9.5 cm) diameter presents.

Normal Delivery

Labour is divided into three stages. The first ends at full dilatation, the second on delivery of the child and the third on expulsion of the placenta and membranes.

In normal labour with a vertex presentation the head usually engages in the transverse position. Uterine activity propels the flexed presenting part through the pelvic brim. Wide sutures and large fontanelles allow moulding of the head. Internal rotation occurs at the level of the levator ani muscles followed by extension, crowning and delivery of the head. As the head delivers it rotates back to its original position (restitution). The fetal shoulders and the rest of the body follow the same path through the pelvis rotating anteroposteriorly. External rotation of the fetal head follows. The trunk flexes posteriorly for the anterior shoulder to deliver and then in the opposite direction to release the other shoulder. The remainder of the body slips out easily. Management of the third stage is described later.

Induction of Labour

Indications for induction of labour include severe hypertension, prolonged pregnancy, growth retardation, rhesus incompatibility, diabetes mellitus, fetal abnormalities, intrauterine death and antepartum haemorrhage. Some would induce older mothers and those with previous infertility or poor obstetric history. Induction should be avoided unless the lie is longitudinal, if two or more caesarean sections have been performed or in the presence of a pelvic tumour. Care should be taken in the grand multipara or if the cervix has previously been repaired. The major risks are fetal (iatrogenic prematurity, cord prolapse and lower Apgar scores), maternal (infection) and increased obstetric intervention after a longer labour often requiring more analgesia. The former is minimised by ultrasonic assessment of gestational age in the second trimester. If the ratio of the phospholipid surfactants lecithin to sphingomyelin is greater than 2, the lungs are probably functionally mature and respiratory distress is unlikely. Meconium or blood contamination should be avoided when testing amniotic fluid.

The Bishop score is based on cervical consistency, dilatation, effacement, position and station of the head. It is used to determine whether ripening of the cervix is required. Vaginal PGE2 and F2a reduce the likelihood of the induction – delivery interval being greater than 24 h and decrease the rate of operative delivery. Risks of prostaglandins include uterine hyperstimulation, fetal distress and fetal death, especially in hypoxic fetuses, e.g. growth retardation.

The aim of induction of labour is to mimic the physiological process of spontaneous labour using either artificial rupture of the membranes or oxytocin infusion. Both are successful in two thirds of patients with a favourable cervix. The starting dose of oxytocin is 1–2 mu/min. This is either doubled at regular intervals in an arithmetic fashion using an automatic infusion pump or the dose is titrated in relation to either uterine contractions or intrauterine pressure measurements. Once labour is established the dose of oxytocin is reduced. Amniotomy and early rather than late oxytocin for induction reduces operative delivery and postpartum haemorrhage. Side-effects of oxytocin include iatrogenic fetal hypoxia due to hyperstimulation, neonatal hyperbilirubinaemia and rarely hyponatraemic fits due to water intoxication when large amounts of oxytocin-containing fluid are used.

Analgesia

Psychological preparation for the pain of labour allied with relaxation and controlled breathing techniques are effective in up to a third of patients. Some require additional pain control in some form. Nitrous oxide (50%) and oxygen (50%) (Entonox) is safe and effective, provided it is inhaled at least 30 s before each contraction. Intramuscular opiates, e.g. 50–150 mg pethidine combined with an antiemetic such as promethazine, depress the fetal respiratory system but this can be antagonised by naloxone. They also slow gastric emptying and may induce confusion and both maternal and fetal tachycardia. Opiates are contraindicated in patients taking monoamine oxidase inhibitors. Regional analgesia is primarily epidural by inserting a fine catheter at L3/4. After a test dose, continuous or pulsatile local anaesthetic, e.g. bupivacaine 0.25%, is injected during the first stage of labour. Epidural anaesthesia after 8 cm dilatation reduces pain in the second stage of labour. Scheduled top-ups reduce episodes of severe pain compared with top-ups at maternal request. Epidural anaesthesia has no effect on the rate of cervical dilatation regardless of whether labour is augmented or not, provided maternal hypotension, hypovolaemia and supine caval compression are avoided. Epidural analgesia is indicated when labour is complicated, e.g. by hypertension, and in the management of breech and twin delivery. It is contraindicated when anticoagulants are in use, when coagulation defects, hypovolaemia, infection at the injection site, active neurological disease or reactions to local anaesthetic agents are present. Maternal effects include: (a) increased duration of the second stage of labour by blockade of sensory roots of S 2,3,4 and the motor nerves of the anterior abdominal wall; (b) 6–20-fold increased incidence of rotational forceps; (c) hypotension, which is prevented by preloading with intravenous fluids; (d) urinary retention; (e) reaction to local anaesthetic agents; and (f) rarely dural tap (leakage of cerebrospinal fluid leading to spinal headache) and infection. Epidural analgesia does not increase perinatal morbidity or mortality. Paracervical and pudendal infiltration with local anaesthetic agents, e.g. lignocaine, is supplemented by blockade of the inferior haemorrhoidal and perineal nerves in the skin prior to operative vaginal delivery. Transcutaneous electrical nerve stimulation (TENS) of painful skin areas may be effective when used correctly but is rarely used.

Chapter 16

Abnormal Labour

Prolonged Labour

Active management of labour involves antenatal education to prepare for the emotional stress of labour, regular assessment of progress in labour and adequate analgesia. Diagnosis of labour is based on satisfactory cervical dilatation in the 1 h period following admission. Subsequent assessment should take place every 1–2 h. The personal attention of one nurse provides social and psychological support and reduces the need for augmentation and caesarean section.

There are three forms of abnormal progress:

1. *Prolonged latent phase* (more than 4 h to reach 3 cm in multiparae and more than 6 h in primiparae).
2. *Primary dysfunctional labour* (dilatation of less than 1 cm/h, which occurs in 26% of spontaneous labour in primigravidae and 8% in multiparae).
3. *Secondary arrest*, which occurs in 6% of primiparae and 2% of multiparae; this is more likely to be obstructed labour (due to malposition, OPP or deflection of the head) than inefficient uterine action.

Failure to progress at the anticipated rate (more than 2 h to the right of a standard partogram) indicates the need for artificial rupture of the membranes and if necessary oxytocin.

Prolonged labour is commoner in primigravidae because of inefficient uterine action. The other major causes are CPD and OPP. Prolonged labour is rare in multigravidae and is usually due to malpresentation or obstruction (of fetal origin) rather than inefficient uterine action. CPD is a functional diagnosis based on the efficiency of labour. Therefore it can be diagnosed only retrospectively by a failed trial of labour. This is defined as a well-conducted labour in a patient with potential disproportion to see whether good contractions will bring about flexion, rotation and appropriate moulding to allow vaginal delivery. Features suggesting the possibility of CPD include stature of less than 152 cm, fetal head more than four fifths palpable at term, clinically small pelvis and a history of difficult vaginal delivery. It is important to distinguish CPD from OPP as a cause of prolonged labour. The two conditions may coexist and OPP may lead to CPD, but delay and operative delivery for OPP is usually not recurrent whereas proven CPD is likely to recur. Two types of abnormal uterine activity are described: (a) inefficiency, which is failure of the uterus to contract sufficiently for satisfactory progress, e.g. abnormal cervimetric progress in the absence of CPD or OPP (though both may coexist with inefficiency); and (b) incoordination, which is characterised by tocographic evidence of irregular uterine contractions. Incoordination is often associated with normal progress of labour. In long-standing obstructed labour a retraction ring may develop at the junction of the upper and lower uterine segments. Rarely a constriction ring develops around the fetus, e.g. when oxytocin is used in the presence of an abnormal lie or following internal podalic version. Treatment is with halothane anaesthesia to relax the uterus.

Abnormal Presentation

Malpositions of the Occiput and Malpresentations

Occipitoposterior Position

In 20% of cases the occiput is posterior at the beginning of labour, usually with the sagittal suture pointing to the right. If the fetus is small or the pelvis is capacious, the head can pass the brim and descend without rotation to deliver without significant delay in the direct OPP. A persistent OPP leads to delay in entry through the brim because the largest diameter of the fetal skull (the occipitofrontal) engages the relatively small space between the sacral promontory and the pectineal line. Once the head reaches the pelvic floor it may fail to rotate anteriorly (deep transverse arrest).

Persistent OPP has been attributed to abnormal pelvic shapes, notably the android and the anthropoid in which the anteroposterior diameter is longer than the transverse. This in turn may lead to deflexion of the fetal head. Inefficient uterine action is probably a much commoner cause. Features suggesting the diagnosis include backache during labour, flattening of the abdomen, limbs felt anteriorly, and the back (where the fetal heart is best heard) well round in the flank. The head is not engaged with sinciput and occiput felt at the same level. Vaginally there is a high deflexed head with the anterior fontanelle in the centre of the pelvis. With increasing flexion and forward rotation the posterior fontanelle becomes more central. If the situation is in doubt (because of moulding, caput), an ear should be identified. The membranes may rupture early, increasing the risk of infection and prolapsed cord.

The second stage may be prolonged. In most (75%) cases the occiput rotates anteriorly and delivers easily; in 5% the occiput persists in the posterior position; in 20% anterior rotation begins but deep transverse arrest occurs at the ischial spines.

Unless there is fetal hypoxia or other complications the main treatment in the first stage is to promote uterine contractions with an intravenous infusion of oxytocin. Epidural analgesia should be used together with adequate intravenous fluids. This may also be successful in the second stage, or delivery can be achieved by manual or Kielland's rotation, or use of the ventouse. In current practice, the latter is preferred. Episiotomy should be performed only once vaginal delivery is certain, i.e. after application of instruments. If these lines of treatment fail, or there are other complications (fetal hypoxia, prolapsed cord in second stage) a caesarean section should be performed.

Face Presentation

This occurs once in every 300–600 deliveries. In many cases there is no obvious cause but well-known associations include anencephaly (10%), prematurity (25%), multiple pregnancy, disproportion, loops of cord around the neck and a swelling in the neck, such as goitre or cystic hygroma.

Three quarters of face presentations are mentoanterior. Most cases arise during labour and are often not diagnosed until just before delivery.

On abdominal examination there is a depression between the anterior shoulder and the head prominence. The fetal heart sounds are heard best on the same side as the limbs. Vaginally the mouth, nose and orbits can be felt (care should be taken not to damage the eyes). The mouth and maxillae form corners of a triangle. The mouth can sometimes be confused with the anus (an ultrasound examination will help to distinguish them). Because the biparietal diameter is 7 cm behind the face, descent may be less advanced than is suggested by vaginal examination; this can often be confirmed by abdominal examination.

Most mentoanterior positions deliver with no assistance or low forceps. Mentoposterior positions cannot deliver vaginally, but around half will rotate spontaneously.

A mentoanterior position should be managed expectantly, with assistance in the second stage only if there is delay. A persistent mentoposterior position requires rotation and extraction or, very commonly, caesarean section.

Brow Presentation

This has an incidence of 1 in 1000 deliveries. The causes and abdominal findings are the same as those of face presentation. On vaginal examination the bregma is felt centrally as a hard, rounded prominence. The anterior fontanelle can be identified, but not the nose, mouth or chin. A large caput succedaneum may form, suggesting greater descent than in reality has occurred (the brow may flex to become a vertex or extend further into a face presentation).

Unless the situation corrects spontaneously labour is prolonged. Failure to progress, or other complications, are an indication for caesarean section.

Breech presentation

This occurs in 1 in 50 deliveries. The incidence is 30–40% at 20–25 weeks, 15% at 32 weeks, and 5% at 40 weeks. The commonest association is prematurity. Others are multiple pregnancy, oligo- or polyhydramnios, hydrocephaly, intrauterine death, pelvic tumours and abnormal uterine shape, either congenital or acquired or because of cornual insertion of the placenta. Extended legs may hinder spontaneous version, and this may explain the higher incidence in primigravidae.

Abdominal examination reveals the head at the uterine fundus and the fetal heart is heard above the umbilicus. On vaginal examination the presenting part is high, soft and irregular. In complete breech the feet can be felt. If the presenting part becomes oedematous it can be difficult to distinguish from a face. The diagnosis and some causes can be confirmed by ultrasound. The three types of breech presentation are:

1. *Extended or frank*: this occurs in 60–70% of cases. The legs are fully flexed at the hip and extended at the knee.
2. *Flexed or complete breech*: the hips and knees are both flexed (10%). Prolapse of the cord occurs in 5% of cases.
3. *Footling or knee presentation* (incomplete): presentation of one or both feet (20–30%). Prolapse of the cord is also common.

The danger to the mother is that of the operative delivery that frequently occurs. The danger to the fetus, after excluding the associated conditions, is perinatal mortality and morbidity. With current management, however, the risk should be little greater than that of a vertex. Specific problems include those of delivery, especially of the aftercoming head: intracranial haemorrhage if this is too rapid; hypoxia if it is too slow; medullary coning; cord severance; brachial plexus injury; fractured bones, including the skull; and crush injuries to the kidneys.

Management of Breech Presentation

Breech presentation can be corrected by external cephalic version after 34 weeks. Spontaneous version is likely up to this time. Contraindications include antepartum haemorrhage, uterine scars, ruptured membranes, oligohydramnios, twins and malformations. Relative contraindications include hypertension, growth retardation, anterior placenta, rhesus isoimmunisation, grand multiparity, obesity and previous caesarean section. If the mother is rhesus negative she should be given anti-D Ig (50 μg or more depending on the result of a Kleihauer test). The procedure may fail, or the breech may revert or it may merely anticipate spontaneous version. Because the procedure may be associated with a 1% fetal loss rate, equal to or greater than that of breech delivery, external version is not widely used.

If there are associated conditions indicating fetal risk (pre-eclampsia, growth retardation, poor obstetric history, older mothers, disproportion, preterm labour provided fetal weight is more than 1000 g, etc.) the child should be delivered by caesarean section. A uterine scar does not exclude vaginal delivery.

Vaginal delivery is acceptable if (a) the pelvis is normal in size (anteroposterior diameters of inlet and outlet greater than 11.5 cm as assessed vaginally and by single upright lateral radiographic pelvimetry, (b) there are no bony pelvic abnormalities, (c) the fetus is normal, (d) the fetus weighs 3.5 kg or less, and (e) the fetus has a normally flexed head.

The first stage of labour is not prolonged, and does not vary with the type of breech. The presenting part usually engages with the sacrum anterior. On meeting the pelvic floor the leading hip rotates anteriorly beneath the pubic arch. The child straightens as the anterior hip is born. As the shoulders enter the brim the trunk undergoes external rotation. They then rotate to bring them into the anteroposterior diameter of the outlet. Finally the head enters the pelvis and rotates until the back of the neck becomes fixed under the subpubic arch. The head is then born by flexion. This is the time at which most problems can arise because the largest part of the baby passes through the pelvis last and there is no time for moulding of the skull bones.

The management of the first stage of labour is similar to that of any high risk pregnancy.

Spontaneous delivery should be avoided. Assisted breech delivery has the following features: (a) epidural analgesia, or local anaesthesia; (b) a generous episiotomy; (c) no interference until the nuchal region is seen, unless the shoulders need to be freed by the Lovset manoeuvre; and (d) controlled delivery of the head by the Burns–Marshall manoeuvre, or better still, forceps. Once the mouth appears the airway is cleared allowing extraction of the rest of the head without haste.

Breech extraction may be indicated for fetal distress or prolapsed cord. However, the procedure carries many risks and caesarean section is generally preferable.

Transverse and Oblique Lies

The incidence is 1 in 300. Associations include a lax multiparous abdomen, contracted pelvis, pelvic tumour, placenta praevia, prematurity, polyhydramnios, multiple pregnancy and abnormal uterine shape.

The uterus is broad and asymmetrical. The fetal head is to one side and the pelvic inlet feels empty. On vaginal examination (after exclusion of placenta praevia) the presenting part (the shoulder) is high and difficult to define.

The membranes rupture early in labour. Incoordinate uterine action is common and the cord may prolapse. If nothing is done labour is prolonged, the fetus dies of asphyxia and the uterus ruptures. Rarely there will be spontaneous reversion to a vertex, or expulsion if the fetus is very small.

The first line of management is exclusion of a specific cause. External version followed by rupture of membranes and oxytocin infusion is used by some. However, unless there is rapid and easy correction, caesarean section should be performed.

Compound Presentation

One or more limbs present with the head or the breech, most commonly a hand with the vertex. The commonest cause is prematurity. Others are contracted pelvis, pelvic tumour, polyhydramnios and dead fetus. The main complication is prolapse of the cord.

Monitoring of Fetal Well-being During Labour

Traditional methods of monitoring fetal well-being in labour include observation of meconium stained liquor (MSL) and auscultation of the fetal heart. The disadvantages are that (a) not all labour-related deaths are associated with MSL or FHR abnormalities prior to delivery, (b) auscultation of the fetal heart for 1 min every 15 min during labour allows only 7% of the continuous FHR to be noted, (c) it may be difficult to hear the fetal heart during and immediately after a contraction, (d) there is considerable observer error in recording the FHR, especially when it exceeds 180 bpm.

MSL in a cephalic presentation and/or an abnormal FHR pattern indicates potential fetal distress. The incidence of MSL is 0.5–20% of all births; 30% in high risk pregnancies; and 22% in post-date pregnancies prior to or in early labour and 44% at delivery. The condition is present in 4% of all pregnancies at 28 weeks and 8% at 38 weeks. If there are no fetal heart abnormalities or acidosis, MSL should be considered a normal physiological finding. Fetuses with thin MSL and normal FHR tolerate labour as well as those without MSL. Apgar scores may be low in the presence of thick MSL.

Fetal asphyxia in utero initiates fetal gasping that may result in meconium aspiration. Meconium aspiration whether in utero or in the neonatal period is associated with a 20-fold increase in neonatal morbidity and mortality. The risk is greatest in acidotic neonates with thick MSL.

The fetal heart has intrinsic rhythmicity set at a rate determined by the sino-atrial node. Sympathetic stimulation increases the FHR and parasympathetic stimulation has the opposite effect. Hypoxia, acidosis, hypothermia and fetal movements all alter the FHR. The speed of response varies with the stimulus. Head compression affects the fetal nervous system and causes immediate changes in the FHR (early decelerations). Uterine contractions lead to anoxia with delayed effects in the FHR (late decelerations). Cord compression alters cord blood flow, which changes the fetal blood pressure and heart rate via baroreceptors (variable decelerations).

The FHR is measured externally by Doppler ultrasonography or internally by direct application of a clip to the presenting part. The following features are examined.

1. *Rate* (normally 110–160 bpm). Tachycardia (moderate > 160 bpm, severe > 180 bpm) may be secondary to hypoxia, acidosis, ketosis or maternal pyrexia. It is most serious when combined with decelerations and/or loss of baseline variability. Bradycardia (moderate 100–110 bpm) is due to increased parasympathetic input and the direct effect of hypoxia on the myocardium. Below 100 bpm (severe) the likelihood of distress or heart block is high especially when complicated by loss of baseline variability (and/or decelerations);
2. *Baseline variability*, defined as fluctuation in the FHR (normally 10–25 bpm). Variability may be long term, with a regular undulating trace with a range of 5–15 bpm occurring every 15–30 s (sinusoidal pattern). These fetuses may be anoxic or anaemic. Short-term variability in FHR is subdivided into type 1 (< 5 bpm, silent), type 2 (5–20 bpm, decreased variability), type 3 (10–25 bpm, normal) and type 4 (> 25 bpm, saltatory pattern). Types 1 and 2 are associated with an increase in low Apgar scores and acidosis but fetal sleep and sedatives may have the same effect;
3. *Accelerations*, defined as an increase in the FHR of greater than 10 bpm lasting 20–60 s. These indicate an intact, uncompromised cardiovascular system capable of responding to stress;
4. *Decelerations*, defined as a fall in the FHR of greater than 15 bpm for more than 30 s. These may be:
    (a) early (synchronous with contractions) and associated with head compression;
    (b) late (beginning after the peak with delayed return to the baseline); half of these fetuses are hypoxic; or

(c) variable in form and recovery: if prolonged they may be associated with fetal distress but are often abolished by maternal positional change.

Up to 50% of fetuses with the most ominous FHR pattern (tachycardia or bradycardia with loss of baseline variability and shallow late decelerations) are proved to be hypoxic.

Though continuous electronic fetal heart rate monitoring (EFM) in labour has many theoretical advantages, it does not reduce the perinatal mortality rate. Other disadvantages include patient acceptability, infection, trauma and cost. Randomised controlled studies of EFM versus intermittent auscultation have demonstrated that the continuously monitored patient, whether she is low or high risk, is more likely to be delivered by caesarean section for "fetal distress". The risk of unnecessary operative delivery is reduced by fetal blood sampling as only 10–50% of ominous FHR patterns are associated with fetal acidosis. There is no difference in Apgar scores or mean umbilical blood pH between intensively monitored and intermittently auscultated patients, but in the former there are fewer babies with cord pH values < 7.2 and the incidence of early neonatal convulsions is lower, especially when fetal blood sampling is used. However, no difference is apparent at 1 year follow-up.

Fetal acid–base status is measured on scalp blood obtained directly through an amnioscope. The normal fetal pH is > 7.25; values between 7.20 and 7.25 indicate mild acidosis and should be repeated after 30–60 min; pH < 7.2 indicates significant asphyxia, especially in the presence of MSL. Metabolic acidosis due to accumulation of lactate and respiratory acidosis from rising fetal P_{CO_2} may also occur. Because of limited glycogen reserves these changes are more pronounced in the growth-retarded hypoxic fetus. The risks of fetal blood sampling are maternal and fetal infection and trauma.

Delay in Second Stage

Once full dilatation is diagnosed (at a varying time after the actual event) a primigravida takes twice as long (40 min) as the multigravida to deliver. This is prolonged by epidural anaesthesia because of motor and sensory paralysis. Bearing down should be avoided until the urge to push is felt. This reduces the need for rotational forceps delivery with epidural anaesthesia. Abnormal FHR patterns and low umbilical artery pH (< 7.25) are less frequent in the upright than in the recumbent position. Oxytocin may be used to stimulate uterine contractions in the second stage, typically after 1.5–3 h in a primigravida or 1 h in a multigravida. If this fails, delivery should be assisted by ventouse or forceps. Sustained bearing down in the second stage is associated with abnormal FHR patterns and low Apgar scores. Exhalatory pushing is preferable.

Prolapse of the Cord

This has an incidence of one in 200–300 deliveries. If the membranes are intact the condition is referred to as "presentation"; once the membranes are ruptured it becomes prolapse. Causes include: (a) poor fit of presenting part (malpositions and malpresentations, especially flexed breech); (b) prematurity; (c) multiparity;

(d) operative procedures (amniotomy, manual rotation); (e) cord abnormalities; and (f) polyhydramnios.

The key to management is recognition of risk factors and early admission. The fetal heart should be monitored continuously and vaginal examination should be performed as soon as the membranes rupture.

Treatment is immediate delivery if the child is alive; the heart beat may still be present even if pulsations are not felt in the cord. Unless the cervix is fully dilated, caesarean section is performed. While preparations are made, the woman is placed in Sim's or Trendelenburg position; the presenting part is supported manually and/or by filling the bladder. The overall fetal mortality is 10–15%, less if the child is delivered within 10 min.

Ruptured Uterus

This is rare, accounting for four maternal deaths in the UK in the period 1994–6. Predisposing factors include long-standing obstructed labour, previous caesarean section (especially classical) or cervical surgery and inappropriate use of oxytocin to augment labour. In some cases there are no obvious predisposing features. Symptoms include abdominal pain (usually severe) and vaginal bleeding. As the uterus ruptures, shock develops (fainting, tachycardia, hypotension), the contractions stop and the presenting part can no longer be felt in the pelvis. Fetal distress precedes fetal demise. Treatment is by laparotomy, with repair of the uterus if possible. Transfusion is usually necessary. Subsequent pregnancies should be delivered electively by caesarean section at 38 weeks, though it is possible for the uterus to rupture at any time during the pregnancy.

Impacted Shoulders

Shoulder dystocia occurs in 2% of all deliveries, particularly when the fetus is macrosomic (e.g. diabetes). After delivery of the head, firm posterior traction is applied to deliver the impacted anterior shoulder. The episiotomy is extended. It may be necessary to rotate the shoulders digitally before delivery. Complications include asphyxia and brachial plexus injury (Erb's palsy).

Chapter 17

The Puerperium

This is defined as the first 6 weeks after delivery, during which time the majority of pregnancy-related physiological changes return to normal. A number of complications may occur.

Postpartum Haemorrhage

Primary postpartum haemorrhage is defined as the loss of 400 ml of blood or more, within 24 h of delivery; it occurs in 2–8% of cases. Its causes include (a) trauma to the genital tract; (b) atony of the uterus secondary to factors such as retention of part or all of the placenta, placenta praevia, abruptio placentae, prolonged labour with uterine inertia (especially in grand multiparae), operative delivery (especially if uterine-relaxing anaesthetic agents, e.g. halothane and cyclopropane have been administered), overdistended uterus with an enlarged placenta (e.g. in multiple pregnancy and polyhydramnios) or multiple fibroids. Apart from haemorrhage, characteristic clinical signs include an enlarged uterus that feels soft and boggy and the patient may be in shock, with a rapid thready pulse and low blood pressure.

The intramuscular administration of 1 ml Syntometrine (0.5 mg ergometrine and 5 units of oxytocin (Syntocinon)) with crowning of the head (active management of the second stage of labour) promotes uterine contraction and reduces the incidence of severe postpartum haemorrhage. In at-risk patients (as defined above) or those in whom ergometrine should be avoided (because of the side-effects of hypertension, nausea and vomiting), oxytocin (10 units intravenously) provides the same effect. If oxytocin infusion was in use during the first stage labour, it should be continued for 1–2 h after delivery.

In the emergency situation, an intravenous infusion should be set up, and the bleeding stopped as rapidly as possible, using intravenous ergometrine or oxytocin. If all or part of the placenta is retained, delivery should be attempted when the uterus contracts. If this fails, resuscitation should continue until the patient is fit for evacuation under anaesthesia. Manual separation of the placenta from the uterine wall is usually easy but may become very difficult with placenta accreta.

This occurs most commonly in older multiparae, typically when placenta praevia or a previous caesarean section complicates the pregnancy. When chorionic villi penetrate the myometrium (placenta increta) or the serosa (placenta percreta), hysterectomy is indicated. However, if future fertility is a requirement, the placenta is left in situ (under antibiotic cover) in the hope that complete autolysis will occur.

If the placenta is complete and the uterus well contracted but bleeding continues, evidence of trauma to the genital tract must be sought. Wound (perineal tear and/or episiotomy) or cervical laceration repair with ligation of damaged vessels generally controls the bleeding. If no vaginal or cervical cause for the bleeding can be found and the uterus is well contracted, a lacerated or ruptured uterus is possible (1 in 2500 deliveries). If this is confirmed at laparotomy and cannot be repaired, hysterectomy is essential.

A broad ligament haematoma is suggested by progressive anaemia with signs of internal bleeding and a mass in one or other iliac fossa. On vaginal examination, a boggy swelling in the lateral fornix pushes the uterus contralaterally. This condition is usually self-limiting; treatment is conservative, with blood transfusion and antibiotics. Surgery should be avoided but evacuation of the haematoma and packing may occasionally be necessary, as may ligation of the internal iliac artery. By contrast, a perivaginal or perivulval haematoma should be drained surgically.

Secondary postpartum haemorrhage occurs between 24 h and 6 weeks after delivery, most commonly between the 8th and 14th day. The lochia is replaced by fresh blood and clots. The uterus may be larger than normal. Immediate treatment includes 0.5 mg ergometrine intramuscularly or intravenously and fundus massage. Ultrasound examination may confirm or exclude retained placental tissue. Antibiotics should be given if there is evidence of infection. In the presence of shock or uncontrollable bleeding, the patient should be resuscitated, followed by surgical evacuation of the uterus. The soft puerperal uterus is easily damaged and it should be repaired immediately. If bleeding persists and a coagulation screen is normal, the uterus should be re-explored and packed. Only rarely is hysterectomy necessary. Chorionic villi are often absent on histological examination. A trophoblastic tumour can be excluded by hCG estimation. Haemorrhage immediately after caesarean section is usually due to bleeding from one of the lateral angle vessels of the lower segment incision. These should be secured surgically. Bleeding a few days later suggests infection and/or retained products. If all else fails, ligation of the anterior division of the internal iliac artery may be necessary. This does not preclude a subsequent pregnancy.

Inversion of the uterus describes the condition in which the fundus descends through the body and cervix into the vagina. Incomplete inversion is difficult to diagnose. Predisposing factors include vigorous attempts at expelling the placenta and injudicious cord traction when the uterus is relaxed. Unless diagnosed and treated immediately (manual or hydrostatic replacement), shock will develop and the maternal mortality rate rises to 15%. There is a 30% recurrence rate in subsequent pregnancies.

Infection

Pelvic infection may occur after prolonged rupture of membranes or labour with multiple vaginal examinations. Retained products of conception or organised blood clot act as a nidus for infection by Gram-positive cocci, Gram-negative coliforms (*E. coli*), *Bacteroides*, *Clostridium* or *Chlamydia*. Pyrexia is accompanied by offensive lochia, lower abdominal pain, uterine tenderness and cervical excitation. Pelvic abscesses present with tender fluctuant pelvic swellings. Peritonitis with septicaemia and shock is rare. High vaginal, cervical and urethral swabs as well as blood cultures should be taken prior to commencing antibiotic treatment. Cephadrine, metronidazole and tetracycline are widely used.

Cystitis and less commonly pyelonephritis may occur in those with a previous history or when catheterisation has been performed in labour. Frequency, urgency and dysuria with pyrexia and renal angle tenderness are typical. The infecting organism is cultured from a mid-stream specimen, which should be repeated once antibiotic treatment is complete. Frequent recurrent infections are an indication for cystoscopy and intravenous pyelogram.

Chest infections typically occur in those who smoke, have chronic bronchitis or have had a general anaesthetic. In severe cases, pulmonary embolism and Mendelson's syndrome must be considered. Productive cough, rales and rhonchi are classical. Physiotherapy and antibiotics are the mainstays of treatment.

Other causes of puerperal pyrexia include surgical wound infection, breast abscess and superficial thrombophlebitis.

Other Urinary Tract Problems

Urinary retention (with or without overflow incontinence) may follow assisted delivery, a painful episiotomy or epidural anaesthesia. An indwelling catheter with continuous drainage for 48 h generally resolves the problem, though this may need to be repeated. The possibility that urinary incontinence may be due to a fistula should be excluded by a three swab test with methylene blue in the bladder. Instrumental bladder injury (e.g. by forceps) typically causes incontinence soon after delivery. Damage following neglected obstructed labour, or, more commonly in the Western world, ureteric damage during caesarean section (leading to bladder sloughing), occurs 8–12 days after delivery. Continuous bladder drainage may allow a small fistula to heal. Delayed surgical repair is otherwise indicated.

Psychiatric Disorders

Puerperal anxiety and depression in the form of the third day blues affect more than half of recently delivered women. No particular hormonal change is responsible. Childbirth is a potent precipitator of severe mental illness: a woman is 20 times more likely to be hospitalised for a psychotic illness in the month after delivery than in any month in the previous 2 years. Between 10% and 15% of mothers have moderately severe postnatal depression between 6 and 12 weeks after delivery, independent of race, culture and social class. Risk factors include: (a) maternal age over 30 years; (b) antecedent subfertility (2 or more years); (c) previous history of psychiatric problems; and (d) life stresses. Clinical features include tearfulness, loss of appetite, energy and libido, insomnia, guilt and self-reproach, isolation and social withdrawal. Tricyclic antidepressants may be indicated. Unlike phenothiazines and benzodiazepines they are safe in lactating women. Psychological support and counselling is essential. Acute psychotic disorders are rare (1 in 500 to 1 in 1000 cases), but serious because of the risk of suicide (10 per million births) and infanticide.

Puerperal psychosis typically occurs within a month of delivery. This may be an affective (manic or depressive) or schizoaffective disorder. Early detection and distinction from postpartum blues is difficult. The majority of patients are primiparae. At special risk are those who have lost a child, or whose child is born handicapped or in need of intensive care. Danger signs include confusion, restlessness, insomnia, hallucinations and delirium. Underlying organic causes, e.g. infection, should be sought. Treatment is the same as that for the disease outside pregnancy. There should be opportunity for the mother to care for her child under supervision. The recurrence rate in subsequent pregnancies is 1 in 5-10; one third have subsequent psychiatric admissions unrelated to childbirth.

Thromboembolism

The new mother has a fivefold increased risk of clot formation (compared with her non-pregnant sister) as a result of pregnancy-related changes in the coagulation and fibrinolytic systems. These return to normal by the end of the puerperium. Blood vessels may be damaged by pelvic surgery or pre-existing varicosities. Bed rest (especially after caesarean section) may reduce blood flow to the lower limbs. Immediately following delivery there is a rise in fibrinogen levels and the platelet count, which is partly countered by the increase in antithrombin activity. Thromboembolic disease (superficial thrombophlebitis, deep venous thrombosis and pulmonary embolism) occurs in 14 per 1000 pregnant and puerperal women. Maternal deaths from pulmonary embolism occur more commonly in the puerperium (caesarean section being a major risk factor). The majority of deaths occur without warning, emphasising the need for preventive measures (early mobilisation and physiotherapy), especially in the high risk group (obese, older women of higher parity, previous deep venous thrombosis or pulmonary embolism during pregnancy or on the contraceptive pill, as well as pre-eclamptic women on bed rest prior to delivery by caesarean section). Prophylactic anticoagulation during pregnancy in women with a previous history of thromboembolism is advocated by some, in an attempt to reduce an estimated 12% recurrence risk. Others prefer to review the evidence upon which the diagnosis was originally made, and reserve prophylaxis for those at especially high risk (bed rest in hospital, two previous thromboembolic episodes or a family history of antithrombin III deficiency). Heparin does not cross the placenta and is the anticoagulant of choice during pregnancy. Long-term use (greater than 6 months) may lead to osteopenia. Increased fetal wastage and bruising/bleeding are other side-effects.

Clinical signs of deep venous thrombosis (oedema, calf tenderness, and positive Homan's sign) are unreliable. To avoid unecessary anticoagulation, venography (using radioopaque dye or isotopes) or Doppler ultrasound must be used in all suspected cases to confirm the diagnosis. Breast feeding is not a contraindication to the use of these techniques, which can identify 85-95% of thrombi.

Large pulmonary emboli may present with acute cardiovascular or respiratory symptoms (chest pain, dyspnoea) and signs (hypotension, cyanosis), but if small these may be absent or transitory. Vigilance and early recourse to chest radiograph, pulmonary angiography or ventilation perfusion scanning are used to confirm the diagnosis. Treatment is by heparin infusion (monitored by the partial thromboplastin time and reversed by protamine sulphate) followed by oral warfarin (monitored by the prothrombin time) for 6 weeks to avoid recurrence. Breast feeding may continue.

Breast Feeding and Breast Problems

Breast feeding has advantages for the nursing mother and her child. It is always available, in a sterile form and at the correct temperature. Lactation protects the infant against infection and atopic illness, inhibits ovarian activity (by altering hypothalamic sensitivity to the feedback effects of ovarian steroids) and encourages uterine involution. Lactational amenorrhoea provides contraceptive protection in up to 90% of women who breast feed fully. Bonding is promoted. Successful breast feeding is encouraged by frequent suckling, correct fixation to the breast, avoiding fluid supplements and providing adequate support (both educational and practical) to the enthusiastic mother.

The major part of breast development occurs at puberty. Breast growth during pregnancy is limited to lobulo-alveolar development and hypertrophy in response to oestrogen, progesterone, human placental lactogen and prolactin. Placental steroids (oestradiol and progesterone) inhibit the action of the lactogenic hormones during pregnancy, such that copious milk production can occur only after delivery. In the absence of suckling, prolactin levels fall to non-pregnant values after 2-3 weeks. Suckling leads to sharp peaks in prolactin secretion, which acts on glandular cells of the breast to stimulate synthesis of milk proteins (casein, lactoglobulin, lactalbumin, lactose) and lipids. This response diminishes over the first 3 months of nursing and is replaced as a stimulus to milk secretion by regular removal of milk from the breast. Release of preformed milk follows the action of oxytocin (released in pulses from the posterior pituitary) on the myoepithelial alveolar cells. This is recognised by the mother as the let-down reflex. Apart from suckling, oxytocin is released in response to visual and auditory stimuli, whereas emotional stress inhibits the milk ejection reflex.

The dopamine agonist bromocriptine (5 mg/day for 14 days) selectively inhibits prolactin secretion and is used to treat breast engorgement in women who elect not to breast feed. Lactation fails in Sheehan's syndrome and in some women taking the combined oral contraceptive pill. Drugs should be avoided unless absolutely necessary.

Human milk is more easily digested than cow's milk; it has a higher energy value, with less protein and major minerals (e.g. sodium) but more fat and lactose than cow's milk. The immunoglobulin (IgA) and white blood cell content provide protection against infective pathogens.

Most nursing women discontinue breast feeding because of fears of having insufficient milk to nourish the infant. More frequent suckling is the answer. Sore or cracked nipples are generally the result of poor fixation to the breast. During growth spurts the infant demands increasing quantities of milk, which should be provided if breast feeding is to succeed. Maternal tiredness necessitates increased domestic support during this period.

A breast abscess appears as a red, painful, fluctuant swelling in a pyrexial, ill woman, typically after the 14th day. Treatment is by surgical drainage and antibiotics. Breast feeding may continue on the opposite breast.

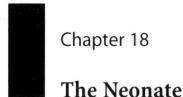

Chapter 18

The Neonate

Examination of the Newborn

Immediately after birth all neonates are examined, usually by a midwife, for gross congenital abnormalities or birth trauma. Premature and small-for-gestational-age infants may need transfer to a Special Care Baby Unit. All babies are re-examined by a paediatrician in a first day check for more subtle abnormalities such as cardiac murmurs, dislocated hips, or dysmorphic features suggestive of various syndromes.

Asphyxia

Birth asphyxia accounts for 7–12% of cases of severe neurological handicap.

Incidence and Definition

There is no standard definition of birth asphyxia, which explains the wide reported range (3–9 per 1000 deliveries). A poor Apgar score (Table 18.1) is the most

Table 18.1. The Apgar scoring table

Feature	0	1	2
Heart rate	Absent	< 100 bpm	> 100 bpm
Respiratory effort	Absent	Slow, irregular	Good, crying
Muscle tone	Limp	Some flexion of extremities	Active motion
Response to stimulation	No response	Grimace	Cry
Colour	Blue, pale	Body pink, extremities blue	Completely pink

The Apgar scoring table is used to determine the condition of all neonates. It is most valuable in identifying those infants in need of resuscitation. It is not a useful way of diagnosing birth asphyxia or hypoxia as it will be affected by other variables, e.g. prematurity, opiate use, meconium aspiration. Scoring is made at 1, 5, and 10 min. bpm, beats per minute.

Table 18.2. Classification of hypoxic–ischaemic encephalopathy

Grade	Clinical features
1	Hyperalert state, tachycardia, jittery, hyperreflexic, dilated pupils, no fits
2	Lethargic but not comatose, bradycardic, hypotonic, poor Moro reflex, miosis, convulsions
3	Flaccid, stuporose, hypotonic, absent suck and Moro reflexes, small mid-position pupils reacting poorly to light, prolonged fits

widely used indicator, though other indicators of asphyxia include a prolonged time to the onset of spontaneous respiration, a low umbilical vessel blood pH, abnormal FHR patterns or abnormal neonatal neurological signs. Except at the extremes of acidosis there remains poor correlation between the Apgar score and acid–base balance.

Though specific to hypoxia, an umbilical artery pH of < 7.10 and a base deficit of > 12 are also used by some to define asphyxia. In low risk groups both indices correlate poorly with subsequent neurological outcomes. Eighty per cent of neurologically abnormal newborns are delivered with a normal cord pH. In a child who has cerebral palsy, a normal pH at birth excludes asphyxia as the cause.

Hypoxic–ischaemic encephalopathy (HIE) is associated with birth asphyxia and is a reliable indicator of future handicap. It is not, however, specific to hypoxia. Birth asphyxia can be ascribed as the cause of subsequent neurodevelopmental handicap only if the HIE is moderate or severe (Table 18.2), irrespective of the cord pH and Apgar score.

Causes and Effects of Asphyxia

Antenatal birth asphyxia may be acute (e.g. abruption and cord prolapse) or chronic (e.g. placental insufficiency). Intrapartum causes include hypertonic uterine contractions, cord compression or prolapse and shoulder dystocia or delay in delivering the aftercoming head of a breech. Hypoxia and acidosis may be present in growth-retarded fetuses, thus decreasing their ability to withstand any additional hypoxic stress even in a normal labour. If the fetus is premature, the risk from birth asphyxia and subsequent handicap is even higher.

Birth asphyxia may produce hypoxia with secondary metabolic acidosis and depletion of glycogen stores. The degree of injury depends on the severity and duration of the asphyxia as well as the gestational age and birthweight of the baby. When severe it may lead to stillbirth or neonatal death. The sequelae of birth asphyxia are legion. The stress may cause the baby to pass meconium prior to the delivery, increasing the risk of meconium aspiration syndrome. This may result in the need for the infant to be intubated and ventilated in order to provide adequate oxygen delivery to the vital organs, and removal of carbon dioxide. In severe cases, newer treatments are now being used such as high frequency oscillatory ventilation (HFOV) with or without inhaled nitric oxide as a dilator of the pulmonary vasculature, or extra-corporal membrane oxygenation (ECMO), a form of lung by-pass therapy. Other sequelae that can result from the asphyxia include renal failure owing to acute tubular necrosis, heart failure, DIC, necrotising enterocolitis, severe acid–base disturbances and cerebral damage.

Clinical Features

The newborn baby may be limp with no spontaneous respiration and bradycardic or even asystolic. Knowing when the last fetal heart beat was detected is of great importance to the paediatrician in attempting to resuscitate such babies. Often, however, the baby will be asymptomatic for the first 12–24 h, after which time the baby has apnoeic episodes that may lead to cyanosis or even convulsions. Raised intracranial pressure may cause periodic breathing, and a vicious circle develops, with worsening cerebral oedema aggravating respiratory function and drive and rising carbon dioxide further increasing the intracranial pressures.

Outcomes

Overall outcomes in babies with HIE depend both on the severity of the insult and the ability to manage the sequelae outlined above. In simple terms, 100% of babies with grade 1 HIE survive with intact neurology. Five per cent of babies with grade 2 will not survive the neonatal period and, of the surviving 95%, approximately 20% will have some form of neurological deficit. Seventy-five per cent of babies with grade 3 HIE will die and 100% of survivors will have neurological deficits, principally spasticity and cognitive disturbances.

Resuscitation of the Newborn

Three out of ten newborns requiring resuscitation come from the apparent low risk groups. Experienced paediatric assistance is essential.

The infant is dried and wrapped in a warm towel. The heart rate and the respiratory effort and rate will determine whether resuscitation is required. In the baby who is not breathing but has a heart rate greater than 100 bpm (primary apnoea), breathing usually occurs spontaneously or after gentle stimulation or facial oxygen. If the heart rate is less than 100 bpm, active measures are taken that may include aspiration of the mouth and nose of secretions, blood or meconium, intubation or bag and mask ventilation with 100% oxygen or the administration of naloxone (0.01 mg/kg) if maternal opiates have been used within the previous 4 h. Failure to respond to these measures should alert staff to check the tube position by listening to the air entry and to assess for adequate chest wall movement, and checking that the pipes from the cylinders are connected, that the cylinders are open and that they are full. Poor breath sounds may be due to a misplaced tube – usually in the oesophagus, a pneumothorax or a diaphragmatic hernia. Asystole or a heart rate less than 60 bpm is an indication for cardiac massage. While this is being done along with adequate ventilation, an umbilical venous catheter should be inserted for the administration of fluid (human albumin, blood, saline or dextrose) and for the rapid administration of drugs (adrenaline 1:10 000, sodium bicarbonate 8.4% or 4.2%).

If meconium had been passed prior to delivery, a suction catheter should be passed either on its own or via an endotracheal tube, and the airways suctioned until no more meconium can be obtained from below the vocal cords prior to the onset of assisted ventilation.

Resuscitation should be stopped if there has been no cardiac activity or output after 10 min, or the infant does not make any spontaneous respiratory effort after 30 min as the likelihood of death or severe quadriplegia is high.

The majority of neonates respond well to prompt resuscitation with greater than 90% of babies with a low Apgar score (0 at 1 min or 0–3 at 5 min) being normal on developmental follow-up.

Specific Problems of the Neonate

Idiopathic Respiratory Distress Syndrome (RDS)

This is a major cause of death in the newborn period and is caused by a lack of surfactant. An estimated 50% of all neonatal deaths relate to RDS or its sequelae and complications. It occurs in 60–80% of babies born at less than 28 weeks' gestation, in 15–30% between 32 and 36 weeks' gestation and in about 5% of babies at term. An increased frequency is found with infants of diabetic mothers, multiple births, precipitous delivery, birth asphyxia, cold stress and a history of affected siblings.

Surfactant is a phospholipid and protein mixture that acts by reducing the surface tension in the alveoli, allowing them to remain open even when at low distending pressures at the end of expiration. When absent, the alveoli tend to collapse at the end of expiration and this leads to respiratory failure.

Clinical features include an expiratory grunt, tachypnoea, chest wall recession and often tachycardia. Cyanosis and oedema may follow. Examination of the chest reveals poor air entry, often with crepitations. The chest radiograph reveals the severity of the disease, ranging from a relatively normal film, through to a hazy so-called ground glass appearance with air bronchograms, to a complete "white out" of the lung fields, with little or no alveolar expansion.

Management involves intubation and ventilation of the infant as determined by arterial blood gas results and the clinical condition of the baby. If the need for ventilation arises, surfactant can be administered directly into the lung via the endotracheal tube. Two doses are often given, usually 12 h apart.

Prognosis depends on the severity of the condition and the availability of expert resuscitation and intensive care facilities. With the advent of newer methods for the management of respiratory failure in this group of babies, overall survival figures are improving for babies of greater than 25 weeks' gestation. Below this age, however, there is still a high mortality and morbidity, often relating to the longer-term problems such as intraventricular haemorrhage, bronchopulmonary dysplasia, necrotising enterocolitis and retinopathy of prematurity.

Idiopathic RDS can be reduced by giving steroids for 48 h to women in preterm labour at less than 34 weeks' gestation. This reduces both the incidence and the severity of RDS by increasing the endogenous production and release of surfactant by the baby and consequently reducing the overall mortality and morbidity.

Jaundice

Neonatal jaundice is common (Table 18.3). It is most commonly an unconjugated hyperbilirubinaemia secondary to an initial failure of the conjugating liver

Table 18.3. Causes, risk factors and clinical features of neonatal jaundice

Disease	Cause	Clinical features	Risk factors
Hepatic immaturity	Temporary deficiency of glucuronyl transferase	Unconjugated bilirubinaemia; preterm: appears after first 24 h, peaks day 4–5. Term: begins 48 h after birth, may last up to 2 weeks	Prematurity
Breast milk jaundice	? Low fluid intake	Unconjugated bilirubinaemia resolves by end of first week	Breast feeding without supplemental fluids
Haemolytic anaemia 1. Acquired (a) Rhesus antibodies (b) ABO and other antibodies	Red cell incompatibility (Haemolytic disease of the newborn)	Unconjugated bilirubinaemia appears within 24 h of birth	Multigravidae; previous transfusion
2. Inherited (a) Hereditary spherocytosis	Fault in red cell membrane	Anaemia and jaundice may be mild, moderate or severe (e.g. congestive cardiac failure, hepatosplenomegaly, intrauterine death due to hydrops fetalis)	Inheritance: (a) autosomal dominant
(b) Glucose-6-phosphate dehydrogenase deficiency	Defect in red cell hexose monophosphate shunt		(b) sex linked
(c) Pyruvate kinase deficiency	Defect in red cell glycolytic pathway		(c) autosomal recessive
(d) Haemoglobinopathies		For details, see table 14.14	(d) autosomal recessive
Infection	Septicaemia Urinary tract infection	Jaundice usually appears after 4th day; mixed conjugated and unconjugated bilirubin	Prolonged ruptured membranes, previous exchange transfusion
Liver disease	Hepatitis Biliary atresia Choledochal cysts	Conjugated bilirubin; jaundice persists beyond first week in otherwise mature infant	
Congenital non-haemolytic jaundice e.g. Crigler–Najjar syndrome	Glucuronyl transferase deficiency	Severe unconjugated hyperbilirubinaemia; kernicterus	Autosomal recessive inheritance
Hypothyroidism		Unconjugated bilirubin; prolonged physiological jaundice	Maternal hypothyroidism and antithyroid drugs, e.g. carbimazole
Metabolic disorders, e.g. galactosaemia	Reduced red cell galactose phosphate	Mixed conjugated and unconjugated bilirubinaemia	Autosomal recessive inheritance

enzyme glucuronyl transferase to switch on at birth. It is a cause of concern because this unconjugated bilirubin is lipid soluble and able to cross the blood–brain barrier. Failure to respond to high levels of unconjugated bilirubin can lead to its deposition in the basal ganglia (kernicterus) and this will result in an athetoid cerebral palsy and or a sensorineural hearing loss. Levels of unconjugated bilirubin have to be very high for this to happen, but in the presence of acidosis, hypoalbuminaemia, hypoxia and sepsis the threshold for damage is also reduced. Once the conjugating enzymes are working the bilirubin can be excreted in the bile.

The level at which an unconjugated serum bilirubin level requires treatment depends on the gestational age, postnatal age and clinical condition of the baby. In mild forms the baby receives phototherapy, which causes isomerisation of the bilirubin in the skin, making it water soluble. In severe cases the baby may require an exchange transfusion with fresh blood compatible with the baby's group and rhesus status.

The major causes of both conjugated and unconjugated hyperbilirubinaemia are shown in Table 18.3. Biliary obstruction due to biliary atresia is an important cause not to be missed. It most often presents as a baby with ongoing jaundice and the passage of pale stools. This condition requires urgent referral to a liver unit for consideration of surgery or liver transplantation.

Hypothermia

The term infants' body temperature falls by 0.5 deg.C after birth. Premature babies are at particular risk of hypothermia because they have a large surface area in relation to body weight, they have less well-developed skin, they cannot shiver and have little or no brown fat stores. Asphyxiated babies are also prone to hypothermia because of malfunction of their normal metabolism.

Hypothermia can cause increased oxygen consumption, vasoconstriction, acidosis, hypoglycaemia, and coagulopathies. It can be prevented by ensuring that the baby is wiped dry as soon as possible after birth and wrapped in warm towels during and after resuscitation. Much of the body heat is lost through the head so a hat or bonnet is important, particularly in the preterm infant.

Small or cold babies are often nursed in an incubator. The temperature is slowly increased as the baby's ability to control its own temperature improves.

Infection

The newborn baby has some immunity from the transplacental passage of maternal IgG but not IgM. This is not usually sufficient to mount an adequate immunological response and this is particularly so for the preterm baby. Infection may be acquired in utero either transplacentally or from infected amniotic fluid. It may occur during vaginal delivery or acquired after birth via the skin, the mucous membranes, the respiratory, urinary or gastrointestinal tracts or haematogenous infection with spread to the brain, meninges, bone or lung.

Clinical features depend on the infection site but are often non-specific, e.g. lethargy, irritability, poor feeding, apnoeic spells, hypothermia or temperature instability. Membrane rupture for greater than 24 h is an indication for close monitoring of the baby. In the presence of any other risk factors (maternal pyrexia, maternal antibiotics, infant pyrexia, prematurity) the baby should have a septic screen (blood culture, urine culture, lumbar puncture and chest radiograph) and be started on antibiotics, most commonly benzylpenicillin and gentamicin.

Overwhelming systemic infections can be fatal, even when treated, though this is rare. Congenital infections from rubella, toxoplasma, cytomegalovirus, syphilis, herpes and HIV may have long-term sequelae (Table 7.4).

Haematological Disorders

Anaemia

Anaemia in the first week of life is diagnosed if the haemoglobin level is less than 12 g/dl. When present it may be due to (a) fetal haemorrhage (from ruptured umbilical cord or placental vessels); (b) haemolysis (acute or chronic) owing to fetomaternal transfusion; (c) twin-to-twin transfusion; or (d) repeated blood sampling. Anaemia after delivery may be due to (a) bleeding, either internally or externally; (b) haemolytic disease of the newborn; or (c) failure of red cell production.

The newborn baby has a circulating volume of approximately 85 ml/kg and hence the rapid loss of large volumes (30–50 ml) of blood results in circulatory failure with shock (pallor, tachycardia and tachypnoea) and a low central venous pressure. If the blood loss is slower the signs may be less obvious until the blood pressure becomes compromised. Once the condition is diagnosed, investigation of the cause must be undertaken. This will include a blood sample for haemoglobin, haematocrit, platelets and reticulocyte count as well as a group and Coombs' test and coagulation screen. A Kleihauer test should be performed on a sample from the mother to look for fetal cells in the maternal circulation.

Emergency treatment includes transfusion (10–20 ml/kg) with O negative blood and the administration of vitamin K (1 mg intramuscularly or intravenously).

Inherited or acquired haemolytic anaemia (Table 18.3) is characterised by reduced red cell survival. These babies may also present with jaundice and their management is outlined above.

Polycythaemia

Polycythaemia is defined as a packed cell volume of greater then 65% and a haemoglobin of greater than 22 g/dl. In twin-to-twin transfusion it may occur in the recipient while the donor becomes relatively anaemic. Delayed cord clamping, intrauterine growth restriction and maternal diabetes predispose to this complication. The baby is usually deep red in colour but if the condition is not detected the baby may present with convulsions secondary to poor cerebral perfusion from the hyperviscous blood, congestive cardiac failure, severe jaundice or respiratory distress.

The management involves a dilutional exchange with plasma or a low saline albumin. The desired end point of the exchange is a packed cell volume (PCV) of 50–55%. This can usually be accomplished with a dilution with 20 ml/kg.

Haemorrhagic Disease

Bleeding disorders may be caused by (a) haemorrhagic disease of the newborn due to deficiency of the vitamin K-dependent clotting factors II, VII, IX and X; (b) neonatal thrombocytopenia due to neonatal platelet antibodies, neonatal leukaemia, maternal SLE or idiopathic thrombocytopenic purpura; or (c) DIC. The baby presents with excessive bruising and bleeding into the skin, at injection sites or into the gut.

Haemorrhagic disease of the newborn is now prevented by the administration of vitamin K (1 mg) to all newborn babies at birth. This is usually by intramuscular injection but is occasionally by three repeated oral doses in the first month.

The other causes are managed according to the aetiology but platelet transfusions are usually required if the platelet count falls below 20 000. The mortality and morbidity from these conditions depend on whether complications such as intracranial haemorrhage develop.

Chapter 19

Obstetric Operations

Amniocentesis

Amniotic fluid is obtained by amniocentesis under ultrasound control. After infiltration of the anterior abdominal wall with local anaesthetic, a fine needle is inserted into the amniotic sac. If the placenta is anterior, it may be necessary to traverse it to reach the amniotic fluid. Amniotic fluid is used (a) for prenatal diagnosis including fetal karyotype, AFP levels and to diagnose metabolic abnormalities after 15 weeks' gestation and later in pregnancy; (b) to diagnose the severity of rhesus disease, and (c) to determine whether the fetal lungs are mature. Complications include a 0.5–1% miscarriage rate and rarely postural deformities, e.g. talipes. Rhesus-negative women should have a Kleihauer test after the procedure and receive 50 μg of anti-D Ig to cover fetomaternal transfusion. Amniocentesis may also be a therapeutic procedure, e.g. intra-amniotic instillation of thyroxine in fetal goitre; repeated drainage amniocentesis in twin-to-twin transfusion.

Chorionic Villus Sampling

Chorionic villus sampling (CVS), best performed at 10–11 weeks' gestation without anaesthesia, provides tissue that permits direct examination of the chromosomal, DNA and metabolic constitution of the fetus. Direct karyotyping is possible within 2 days; formal karyotyping and other studies yield results within 2–3 weeks. The indications for CVS are similar to those for amniocentesis in the first half of pregnancy, the commonest being fetal karyotyping because of advanced maternal age.

CVS is performed via the cervix or the abdomen, the latter being the commonest route. Complications are uncommon (1–2% miscarriage rate; premature rupture of membranes before term) and are closely related to operator experience. CVS performed prior to 8 weeks' gestation may lead to severe limb reduction deformities and other congenital abnormalities. A major problem is mosaicism in cultured placental cells, which may not reflect the chromosome pattern of the fetus; this problem is avoided by early amniocentesis.

Fetoscopy and Cordocentesis

Percutaneous transabdominal intrauterine endoscopy may be performed at 15–18 weeks' gestation to diagnose or exclude an anatomical abnormality under direct vision or to obtain tissue or fluids (e.g. cord blood) for analysis. The indications for fetoscopy alone are diminishing and in most cases the diagnosis can be made by ultrasound. This also allows collection of fetal blood (cordocentesis) and other needle-guided procedures, e.g. fetal intravascular transfusion or doing administration as well as drainage procedures of fetal hydrothorax, abdominal cysts and bladder. Selective fetal reduction is best performed at 11 weeks' gestation as the miscarriage rate is lowest (10%) at this time.

Termination of Pregnancy

In the first trimester, termination of pregnancy may be performed medically up to 8 weeks' gestation (the synthetic prostaglandin analogue Gemeprost and the antiprogesterone mifepristone) or surgically by dilatation of the cervix (to no more than 10 mm) and suction evacuation using a narrow suction cannula. Blood loss is reduced by postoperative ergometrine (0.5 mg intravenously). Second trimester termination is most often performed by extra-or intra-amniotic prostaglandins (PGE2 or F2a). Sharp curettage is not recommended in the second trimester because of the increased risk of uterine perforation, haemorrhage, infection, cervical laceration and incomplete emptying of the uterus. Rarely, hysterotomy is performed via a low transverse incision. The uterovesical peritoneum is opened and the bladder reflected downwards. A short vertical incision is made as low as possible in the uterus. The bulging membranes are separated from the inner surface of the uterine cavity and the sac contents removed. The uterine incision is closed in two layers and covered by the uterovesical peritoneum.

Complications of termination of pregnancy include: (a) haemorrhage precipitated by retained products or endometritis; (b) endometritis that may lead to tubal damage and infertility; (c) damage to the cervix and uterus including cervical incompetence, perforation and subsequent uterine rupture; (d) infertility secondary to endometritis or Asherman's syndrome (intrauterine adhesions); (d) rhesus isoimmunisation; and (f) psychological problems.

Sterilisation

Female sterilisation is performed as an open technique at (mini) laparotomy, via laparoscopy or vaginally through the pouch of Douglas, preferably in the first half of the menstrual cycle to minimise the possibility of coincident pregnancy. The fallopian tube is ligated (Madlener technique) and/or divided (Pomeroy technique). The tubal stumps may be buried in the broad ligament. Clips (Filshie, Hulka–Clemens) or rings may be applied transabdominally, or intratubal devices inserted transcervically. Sections of the tube may be destroyed by diathermy (unipolar or bipolar) but this causes extensive damage and reduces the likelihood of successful reversal. Failure rates vary with the technique and are approximately 0.2% (Filshie clip), 2.6% (Hulka–Clemens clip) and 4% (Madlener technique, Yoon ring). Clip failure may be due to application to the wrong structure, recanalisation or fracture of the clip. Apart from the anaesthetic risks and those of the procedure (mortality 8 per 100 000 laparoscopies), complications include pregnancy (including a 5–8% ectopic pregnancy rate), injury to the bowel, menstrual irregularity and loss of libido. The risks, failure rate and likelihood of reversal (approximately 50–70% depending on the procedure) is included in the counselling and written consent that precedes sterilisation.

Vasectomy involves excising a segment of the vas deferens, usually under local anaesthesia. Sterilisation is complete when no spermatozoa are found in the ejaculate. Reversal is difficult and even when surgically possible half the men will be infertile because of sperm antibodies. Failure may be due to insufficient excision or spontaneous recanalisation. Complications include infection (epididymo-orchitis), granuloma and haematoma formation.

Cervical Suture

Cervical incompetence may be congenital or iatrogenic, e.g. due to overvigorous dilatation, cone biopsy or cervical amputation. A history of cervical incompetence suggested by previous mid-trimester abortions or recurrent premature delivery is an indication for cerclage (surgical closure of the cervical canal in pregnancy). When performed between 12 and 14 weeks, cervical cerclage reduces the incidence of delivery before 33 weeks, miscarriage and perinatal death. The McDonald procedure consists of taking three bites of the cervix with a large needle threaded with Mersilene tape. The suture is tightened to just close the os. Prophylactic antibiotics and beta sympathomimetics are used by some. Cerclage with colpotomy (Shirodkar's procedure) requires dissecting the bladder to the level of the internal os prior to inserting the purse string suture of non-resorbable material. The suture is removed before labour is established or when the membranes rupture or haemorrhage occurs. Transabdominal insertion of a cervical suture may be indicated if gross anatomical distortion prevents the insertion of a suture transvaginally. Complication rates are low and variable and include cervical laceration, premature rupture of membranes, sepsis, cervical stenosis and dystocia, fistula and uterine rupture.

Version for Breech Presentation

Version may be external or internal. External cephalic version is used by some to convert breech into cephalic during the third trimester. The fetus is rotated by gentle pressure on the anterior abdominal wall. Necessary conditions include a relaxed uterus, ample liquor and a disengaged presenting part. Anaesthesia should not be used. External version can also be used in early labour with a transverse or oblique lie; the membranes should be intact and placenta praevia, pelvic tumour and disproportion must be excluded. When delivery of a second twin needs to be expedited, external version may be used prior to rupturing the membranes and applying the ventouse. Rarely, internal podalic version is used to extract a second twin with a malpresentation. Risks include rupture of the uterus and fetal death.

Caesarean Section

There are few absolute indications for caesarean section. It is performed either electively (e.g. for fulminating pre-eclampsia, growth retardation, prematurity, confirmed disproportion, two previous caesarean sections, or when vaginal delivery is likely to be more risky to the fetus and/or mother, e.g. placenta praevia) or during labour for a variety of reasons including fetal or maternal distress and evidence of obstructed labour. The overall incidence is 10-25% of all deliveries. A policy of trial of labour after previous caesarean section reduces maternal morbidity and the incidence of repeat caesarean section. Intrauterine pressure should be monitored if oxytocin is used after a previous caesarean section. The maternal mortality of emergency caesarean section is 10-fold greater than that of vaginal delivery, though the rate has declined steadily in recent years. The major immediate causes of death are (a) pulmonary embolism, (b) hypertensive disease and (c) haemorrhage.

Most sections are performed transversly through the lower segment after reflecting the bladder. Classical caesarean section is a vertical incision through the upper uterine segment. It is performed when entry through the lower segment is obstructed by fibroids or adhesions or for a transverse lie with premature rupture of membranes, especially at 26-28 weeks' gestation. Procedures such as classical caesarean section, previous hysterotomy or myomectomy when the uterine cavity was entered are associated with an increased risk of scar rupture in subsequent pregnancies, especially if sepsis followed the operation or the placenta implants over the scar. The maternal risks are those of any operation but in addition the bladder, ureter or bowel may be injured. Prophylactic antibiotics reduce the incidence of serious wound infection and febrile morbidity in elective and emergency caesarean section.

Forceps Delivery

Forceps delivery is considered for fetal or maternal distress (including some conditions when it is preferable to avoid pushing) and for delay in the second stage due to poor maternal effort, malposition or malpresentation. Forceps are also applied to protect the aftercoming head in a breech presentation. However, forceps do not protect the low birthweight infant from intraventricular haemorrhage.

Forceps should be applied only when (a) the cervix is fully dilated, (b) the head is engaged with a vertex (anterior, lateral or posterior position) or face presentation (mentoanterior position), (c) the biparietal diameter is at or below the level of the spines, (d) the membranes are ruptured, and (e) the bladder is empty. Significant caput or moulding of the head may obscure these observations. Regional epidural or pudendal and inferior haemorrhoidal local anaesthesia is required.

If the vertex presents in the posterior position it may first be rotated either manually or with Kielland's forceps before delivery unless it is deep in the pelvis, when delivery in the posterior position is preferable. Keilland's forceps have a sliding lock and are used to rotate the head from the posterior or transverse position when the vertex is at or below the level of the spines. The forceps are applied either directly or by wandering the anterior blade around the face onto the anterior parietal bone. This is the only type of forceps that does not have both pelvic and cephalic curves. Straight forceps in common use include Simpson's, Neville–Barnes, Anderson's and the short handled Wrigley's. Each has a fixed lock between the blades. Intermittent traction is applied during contractions to deliver the head from the right to left oblique or the direct OPP. Forceps delivery may fail because of an unrecognised malpresentation, disproportion or hydrocephalus, but more usually because of faulty technique. Injury to the mother (uterine rupture, vaginal, bladder or bowel laceration) or fetus (facial abrasion, bruising, or nerve palsy and intracranial haemorrhage) may result.

Ventouse Delivery

The incidence of ventouse delivery is now higher than forceps delivery in the UK. The indications and precautions are as for forceps delivery, except that the ventouse may be used when the cervix is not fully dilated and local anaesthesia is all that is usually required. It should be avoided in the preterm infant. The largest possible cup is applied as far back on the occiput as possible and the pressure increased steadily to 0.8 kg/cm^2. Traction is exerted at right angles to the cup during contractions for a maximum of 30 min to minimise the risk of scalp trauma (bruising, sloughing or cephalhaematoma). Transverse or posterior positions of the head may rotate on the pelvic floor and deliver occipitoanteriorly. Maternal injury is rare. For all indications the incidence of maternal injury and use of general anaesthesia is less than with forceps delivery.

Episiotomy and Repair (Including Tears)

An episiotomy is performed electively (after local anaesthesia) in many labours to extend the anteroposterior space available for delivery of the head. It is an incision through the perineal body extending either in the midline posteriorly or mediolaterally through the vulva and vagina. Overall trauma and perineal trauma are reduced by a policy of restricted rather than liberal use of episiotomy. Three types of tear are described: first degree involving perineal skin and the lower part of the vagina; second degree involving the perineal body, levator ani muscles and posterior vaginal wall; and third degree extending through the external anal sphincter into the anus or rectum.

A third degree tear is repaired in an operating theatre. Interrupted sutures are placed starting at the apex of the anal mucosa down to the anal margin, followed by the torn ends of the sphincter. Repair of an episiotomy and a first or second degree tear is in layers using either continuous or interrupted polyglycolic acid (absorbable) sutures starting at the apex of the vagina down to the introitus. Two layers of interrupted sutures close the dead space behind the vagina and repair the perineal body. Finally the skin is closed, usually with continuous polyglycolic acid sutures. Interrupted catgut, silk and nylon sutures increase short-term perineal pain. Other complications include bleeding, urinary retention and difficulty with defaecation due to pain, infection that often leads to wound breakdown, perivaginal or perivulval haematoma and dyspareunia. Most are reduced by expert repair.

Manual Removal of Placenta

Indications and complications are covered elsewhere. If the placenta has separated, it may be retained by a retraction ring between the upper and lower uterine segments. If there is no active bleeding, the uterus may be allowed time to relax; delivery of the placenta follows. Oxytocin infusion is continued after the placenta is delivered. If the retained placenta has not separated, it is removed by digital separation through the spongy layer of the decidua basalis, usually under general anaesthesia. Manual separation of the placenta from the uterine wall is usually easy except with placenta accreta. Hysterectomy is indicated when chorionic villi penetrate the myometrium (placenta increta) or the serosa (placenta percreta). If future fertility is a requirement the placenta may be left in situ under antibiotic cover.

Section II
Gynaecology

Chapter 20

Menstrual Disorders

Puberty

The onset of puberty is under pineal (melatonin) and hypothalamic (gonadotrophin-releasing hormone (GnRH)) control (see companion volume, *Basic Sciences for Obstretics and Gynaecology*, Chard and Lilford, 1995).

Most (95%) girls menstruate between 11 and 15 years of age, the remainder doing so between 10–11 and 15–16 years. Improved nutritional and environmental factors have been responsible for the decline in the age of menarche this century. This now appears to have stabilised. Menarche is typically the last stage in pubertal development, preceded by growth in height, breast development and the appearance of pubic and axillary hair. Frequent variations in this pattern occur.

Precocious puberty is diagnosed if there is breast and pubic hair growth before the child is 8 years old, or menstrual periods before the age of 10. Most cases are constitutional, i.e. premature gonadotrophin release, in the absence of organic pathology, e.g. intracranial lesions (tumour, infection), Albright's syndrome or a feminising ovarian tumour. In the absence of organic pathology, menstruation can be suppressed with cyproterone acetate (70–150 mg daily). Most feminising tumours are benign and oophorectomy is sufficient.

Most early menstrual cycles are anovular and hence irregular and pain free. If periods are heavy or frequent, anaemia may result. Norethisterone (5 mg three times per day) will stop the bleeding. Rarely curettage is necessary. The possibility of pregnancy and its complications as well as drug intake (e.g. oral contraceptive) should be considered in the differential diagnosis of bleeding disorders.

Menorrhagia

The amount, duration and interval between periods vary considerably among normal women. The diagnosis of menorrhagia is based on the patient's own subjective assessment. Two thirds of women complaining of menorrhagia have objective evidence of heavy menstrual blood loss (> 80 ml per cycle) as do 10% of a random population of women. Organic disease of the genital tract is excluded

Table 20.1. Causes of menorrhagia. (Note that histological examination may reveal unsuspected organic pathology, e.g. endometritis, polyps, malignant disease in up to 25% of patients with dysfunctional uterine bleeding in whom pelvic examination fails to detect abnormalities

Organic disease of the genital tract		Dysfunctional uterine bleeding
Vulva:	Polyps	**Primary:** Unknown; probably multifactorial disturbance of endometrial function
Vagina:	Polyps, adenosis	
Cervix:	Polyps, carcinoma	**Secondary:** Disorders outside the genital tract (myxoedema, bleeding complications, polyps, malignancy, bleeding diathesis, drugs including contraceptives), endometritis (e.g. TB)
Tube:	PID, pregnancy complications	
Ovary:	Hormone producing tumours	

PID, pelvic inflammatory disease; TB, tuberculosis.

by general and vaginal examinations followed by transvaginal ultrasound, endometrial biopsy or hysteroscopy. There is no pattern of bleeding (oligomenorrhoea, polymenorrhoea, menorrhagia, metrorrhagia) specific to any disorder listed in Table 20.1.

Mechanisms of Menstruation

Menstrual bleeding may be secondary to a fall in oestrogen levels, or to decreased progesterone (P) levels in an endometrium primed by oestrogen (E) (normal cycle). Progesterone is not essential for withdrawal bleeding to occur; anovular cycles also result in endometrial shedding. The effects of E and P on the endometrium are mediated by the prostanoids (PGE2, PGF2a, thromboxane and prostacyclins). Oestrogen depresses and progesterone stimulates activity of the initial enzyme in the degradation of PGE2 and F2a (15-hydroxyprostaglandin dehydrogenase). The net result of the fall in progesterone levels prior to menstruation is release of phospholipase A2 (an acyl hydrolase), initiating the arachidonic acid cascade. Prostanoids thus increase during the secretory phase reaching a maximum premenstrually.

Prostanoid levels are higher in patients with menorrhagia (including intrauterine contraceptive device (IUCD)-related cases), endometriosis, dysmenorrhoea and endometrial carcinoma. PGE2 levels (vasodilator) are higher than PGF2a (vasoconstrictor) in patients with menorrhagia. Prostacyclin (vasodilator and antiplatelet aggregator) but not thromboxane (vasoconstrictor and platelet aggregator) metabolites are increased in menorrhagia. These observations form the basis for the therapeutic success of prostaglandin inhibitors in reducing menstrual blood loss.

Dysfunctional Uterine Bleeding (DUB)

DUB (abnormal uterine bleeding in the absence of organic genital tract disease), affects 10% of all new gynaecology outpatients. Half of them are aged 20–40 years, 40% are over 40 and 10% are less than 20 years old. Contrary to common belief, no pituitary–ovarian abnormality (or anovulation) is present in the

Table 20.2. Clinical types of abnormal uterine bleeding

Age	Frequency (%)	Type	Features	Prognosis
< 20	10	Physiological	Transient irregular or light periods	Generally good; worse if abnormal bleeding starts at menarche than after a period of normal menses
		Pathological	Persistent and/or severe e.g. TB, endometritis	
20–40	50	Cyclical	Usually ovulatory (irregular endometrial ripening or shedding)	Good; most remit spontaneously
		Acyclical	Usually organic disease, anovulation (proliferative or hyperplasia)	Depends on diagnosis, may recur
> 40	40	Usually acyclical or intermenstrual	Perimenopausal change in pituitary–gonadal function Serious organic disease (uterine cancer)	Depends on diagnosis

TB, tuberculosis.

majority of women. The primary cause of the menstrual dysfunction is unknown. The traditional classification of DUB according to aetiology (ovulatory, anovulatory and corpus luteum abnormality) or complex nomenclature (e.g. menometrorrhagia) is being replaced by one based on clinical presentation (affected age groups and bleeding pattern) supported by histology (Table 20.2). Women with DUB who remain unresponsive to therapy should be further investigated, e.g. hysteroscopy and laparoscopy.

Pathology

Most cases are best sampled 5–6 days before menstruation. Endometrial histology is normal in up to half the women with DUB. Approximately 30–40% have endometrial hyperplasia (metropathia haemorrhagica) with overgrowth of stroma and glands (Swiss cheese appearance) or proliferative endometrium in the second half of the cycle (anovulatory). Atrophy, irregular ripening and irregular shedding of the endometrium (ovulatory) constitute the remaining histological types seen less commonly in DUB.

Treatment

Treatment is either conservative (drugs) or radical (endoscopic endometrial ablation or resection, hysterectomy). Surgery is limited to older women whose families are complete and in whom persistent bleeding has been unresponsive to medical treatment. In women desirous of pregnancy or when normal secretory endometrium is present, prostaglandin synthetase inhibitors (flufenamic and mefenamic acid) with or without antifibrinolytic drugs (epsilon aminocaproic acid, tranexamic acid) or the progesterone-loaded intrauterine contraceptive device are the treatment of choice. Menorrhagia may recur once treatment is stopped in some women, though the remainder are "cured". If contraception is desired and/or atrophic endometrium is present, the first choice is an oestrogen-dominant combined E/P oral contraceptive, given from days 5 to 25 for 3 months in the first instance. Normal menses follow in the majority of women due to a rebound effect that normalises pituitary–ovarian–endometrial axis function. If secretory phase endometrial sampling demonstrates proliferation, a progesterone-dominant E/P combination pill should be used; if hyperplasia is diagnosed, cyclical progesterone therapy should be tried for 9–12 months. Some advocate hysterectomy if the condition persists because of the small risk of progression to invasive carcinoma. The progestogenic drug danazol (a 17α ethinyl-testosterone derivative) is more effective in reducing menstrual blood loss than the E/P combination pill, as are the GnRH analogues but their use is limited by expense and unpleasant side-effects (weight loss, rashes, irreversible hoarseness (E/P/ pill) and menopausal symptoms (GnRH analogues)).

Primary Amenorrhoea

The single most common cause of primary amenorrhoea is congenital absence of the internal genitalia (the Mayer–Rokitanski–Kusler–Hauser syndrome). The 15 year old without periods or secondary sexual characteristics should be investigated, though, if the latter are present, investigations can be deferred. Apart from a full history and physical examination (including height), follicle-stimulating hormone/luteinising hormone (FSH/LH) assay, chromosomal studies and a skull radiograph are performed.

Androgen Insensitivity Syndrome

In androgen insensitivity (of which testicular feminisation syndrome is one subgroup), XY "females" with testes present at puberty with normal breasts but absent pubic or axillary hair and primary amenorrhoea. The finding of a blind short vagina should lead to investigation of the internal genitalia. The hypothalamus and pituitary are insensitive to testosterone, which is thought to be responsible for the elavated LH values. Insensitivity to androgens is usually due to failure of receptor binding. Male levels of E, FSH and testosterone are found. Inheritance is by an X-linked recessive or sex-limited autosomal dominant gene. Depending on when the diagnosis is made (typically at puberty, but occasionally earlier when the same diagnosis is made in an elder "sister"), gonadectomy should be carried out, because the testes carry a 5% lifetime risk of malignant change. Hormone replacement therapy should follow at puberty.

Congenital Adrenal Hyperplasia

This should be distinguished from prenatal drug-induced masculinisation (danazol, methyltestosterone, E/P) which typically presents during infancy. Late-onset congenital adrenal hyperplasia may present at puberty in two ways. First, the female with minimal masculinisation of the external genitalia may present with delayed secondary sexual development and amenorrhoea. A number of enzyme defects of cortisol and aldosterone metabolism are possible, the commonest being 21-hydroxylase deficiency. Elevated serum levels of 17αhydroxyprogesterone, or of its urinary metabolite pregnanetriol, will confirm the diagnosis. Urinary androgens (17α-oxosteroids) are also elevated. The associated aldosterone deficiency responsible for the high mortality of this condition at birth (salt-losing syndrome) is rarely a problem at puberty. Treatment with cortisol breaks the positive feedback cycle on pituitary adrenocorticotrophic hormone (ACTH), thus removing the stimulus for adrenal androgen production. Secondary sexual development and menstruation result. Second, the diagnosis may be made in a "boy", so identified because of severe masculinisation of the external genitalia at birth. Cortisol treatment will induce menses. The decision to change the gender role should not be taken lightly, especially if the penis is thought to be suitable for intercourse. In this case it may be better to allow the child to continue male sexual development by treating with testosterone and performing hysterectomy and oophorectomy.

True Hermaphroditism

The "female" hermaphrodite will become virilised at puberty unless the testis is removed. Similarly, menstruation and breast development may occur in the "male" hermaphrodite at puberty. The distinction between an undermasculinised male (due to testicular failure, whether for anatomic or enzymatic reasons, or end organ hormone insensitivity) and a true hermaphrodite (possessing both ovarian and testicular tissue) can be made only by gonadal biopsy at laparotomy, since it is difficult to distinguish either of these gonads from an ovotestis at laparoscopy. When gonadal tissue inappropriate to the chosen gender role is found (based on the suitability of the external genitalia for sexual life in that role) it should be removed to reduce the lifetime risk of malignancy. Once removed, E/P treatment at puberty will produce secondary sexual development and menses in those who are best suited to be female, whereas testosterone will achieve the opposite effect. Occasionally this culminates in fertility.

Hypergonadotrophic Hypogonadism

Sexual maturation in Turner's syndrome and gonadal agenesis can be achieved with oestrogen and withdrawal bleeding induced with added cyclic progesterone. The gonadal streak should be removed in all XY or XX/XY mosaics to prevent the risk of a dysgenetic tumour developing. Assisted reproductive techniques (ovum donation) may be used in patients with gonadal agenesis and a normal uterus.

The Undermasculinised Male

Genetic males with enzymatic faults in androgen production (e.g. 5α-reductase deficiency) may present either in infancy when an elder "sister" is being investigated for the same condition, or at puberty when the "girl" fails to menstruate, and varying degrees of virilisation are induced by testosterone. The internal genitalia are male since anti-müllerian hormone is produced normally. Inheritance is autosomal recessive. Typically the female role is preferred, since the external genitalia are barely adequate for male sexual activity.

Secondary Amenorrhoea

This is defined as the absence of menses in women of reproductive age who are not pregnant, lactating or have had a hysterectomy. The causes are physiological (weight loss, stress, exercise), pharmacological (oral contraceptive pill, drugs that induce hyperprolactinaemia such as phenothiazines, tricyclics, reserpine, digoxin) and pathological (asymptomatic: pituitary tumour, hypothyroidism; following curettage: Asherman's syndrome; following postpartum haemorrhage: Sheehan's syndrome; with hirsutism: adrenal disease, polycystic ovaries, virilising ovarian tumours; with galactorrhoea: pituitary tumour; with hot flushes or following radiotherapy or chemotherapy: premature ovarian failure). Accurate diagnosis (which should always precede treatment) is based on adequate history and physical examination (see above). Organic disease should be excluded in the first instance by FSH and prolactin assay, as well as skull and pituitary fossa radiographs. If fertility is not important to the patient, no further investigation is necessary. The management of the amenorrhoeic woman wanting pregnancy is discussed in a later section.

Hyperprolactinaemia

Stress (including vaginal examination) increases prolactin levels. Hyperprolactinaemia per se induces secondary amenorrhoea, which is treated with bromocriptine (side-effects include nausea, vomiting, postural hypotension and headaches). Pituitary tumours may cause amenorrhoea without hyperprolactinaemia but this is rare. Hyperprolactinaemia (with or without galactorrhoea) and amenorrhoea are an unusual (4%) presentation of hypothyroidism.

Polycystic Ovarian Disease

This includes secondary amenorrhoea, oligomenorrhoea, obesity, hirsutism and infertility. Persistently high LH levels and a high LH/FSH ratio are suspect, while enlarged polycystic ovaries at laparoscopy or ultrasound are conclusive. Treatment of infertility is by clomiphene citrate (CC) or human menopausal gonadotrophins (hMG) or recombinant FSH (rh FSH) to induce ovulation. In CC-resistant women (i.e. no ovulation after up to 200 mg CC daily for 5 days) surgery by ovarian endoscopic electrocautery or laser "drilling" is preferred. Clomiphene is an anti-oestrogen resulting in an increase in FSH/LH. Side-effects include hot flushes and, in 7% of cycles, asymptomatic ovarian enlargement. Regression follows cessation of treatment. There is a 10% incidence of multiple pregnancy, but these are rarely of higher order than twins (unlike hMG, which induces multiple pregnancy in 20% of conceptions, including a 3% triplet rate). Endoscopic surgery to the ovaries produces fewer adhesions and leads to spontaneous ovulation in 60% of women.

Pituitary Tumours

Radiographic evidence of ballooning of the fossa, blistering of the floor or erosion of the clinoid processes is pathognomonic. Contrast CT scanning is the only reliable method to exclude pituitary microadenoma. These may recur despite transsphenoidal excision; bromocriptine is the treatment of choice. By contrast, mixed tumours or craniopharyngiomas (which cannot be excluded when prolactin levels are raised) are best treated surgically.

Premature Menopause

Elevated FSH/LH associated with low oestrogen levels suggest premature ovarian failure (in women below the age of 35 years) or rarely the resistant ovary syndrome. Open ovarian biopsy will reveal atretic Graafian follicles in the former condition. Tissue antibody screening identifies a subgroup of women with premature menopause at risk of developing other autoimmune diseases.

Dysmenorrhoea

Primary spasmodic dysmenorrhoea is typical of the young woman when regular ovulation is established. The pain is generally short lived (rarely longer than 24 h) and colicky in nature. Treatment depends on the severity (i.e. the extent with which it interferes with school, work, etc.). The aetiology is unknown, though the effectiveness of aspirin and prostaglandin synthetase inhibitors suggests a role for the arachidonic acid cascade. By inhibiting ovulation, the oral contraceptive pill is generally effective.

Secondary congestive dysmenorrhea typically starts a few days before the menses are due and continues through most of the period. The differential diagnosis is between endometriosis (with dyspareunia, pelvic and low back pain, menorrhagia and/or irregular menses) and chronic pelvic inflammatory disease (similar but with chronic cervicitis). Laparosopy distinguishes between the two.

The Premenstrual Syndrome

The premenstrual syndrome is a complex of physical and psychological symptoms of unknown aetiology, not associated with organic disease, which recur regularly during the same phase of each ovarian cycle and then regress. Ovulation but not menstruation is essential to the diagnosis. The types of symptom (physical (e.g. oedema, weight gain, bloating and breast tenderness) and psychological (e.g. irritability, tiredness, depression)) but not their severity, remain constant from one cycle to the next. The diagnosis is usually based on history and self-assessment questionnaires. Menstrual charts are helpful in defining the problem and monitoring the treatment. The prevalence among women of reproductive age ranges from 20% to 40%, though it is ten times lower on the basis of strict diagnostic criteria. It occurs most frequently in women over 30 years, increasing in severity towards menopause. Symptoms begin 5–14 days before menstruation, subsiding 1–2 days after.

Sympathetic handling is important. There is a placebo effect of up to 75% with any treatment. Many women prefer to control their symptoms by altering their diet (eating small, frequent protein rich meals and limiting salt, caffeine, animal fats, and refined sugars), exercise especially in bright light and reducing stress.

Treatment is symptomatic. Options include (a) pyridoxine (100 mg daily) for depression and tiredness or gamma-linolenic acid for mastalgia; (b) progestogens (e.g. progesterone suppositories 200–600 mg) from mid-cycle onwards; (c) diuretics (e.g. spironolactone 50 mg, or bendrofluazide 2.5 mg daily) for bloating; and (d) prostaglandin inhibitors, e.g. mefenamic acid or naproxen for menorrhagia and dysmenorrhea. More serious cases may require suppression of the ovarian cycle with the oral contraceptive pill, oestradiol (patches or subcutaneous implants), danazol or GnRH agonists, anxiolytics or antidepressants. Of these, only GnRH agonists, alprazolam and serotonin-reuptake inhibitors have been shown to be effective in prospective, double-blind, placebo-controlled trials.

Dyspareunia

Dyspareunia is pain on intercourse, described as superficial when the pain is at the introitus and deep when it is felt within the pelvis. It may also be defined as primary if present since first attempts at intercourse and secondary if it has arisen later. In many cases no organic cause is found but in some there are specific conditions such as infections, pelvic inflammatory disease, tumours, endometriosis or a fixed retroverted uterus.

Chapter 21

Menopause

Menopause is the cessation of menses. The climacteric (perimenopause) is the 10–20 year period around this point. The median age is 50 years, irrespective of race, socioeconomic status, parity, height or weight. In smokers it occurs 1–2 years earlier. It is due to a decline in the supply of ovarian follicles (oocytes) by atresia, a process which begins after the 20th week of intrauterine life.

Stages of the Climacteric

The process begins at the age of 35–40 years (compensated ovarian failure). There is increased resistance of the ovarian follicles to gonadotrophin stimulation. Consequently, follicular development is deficient, 25% of cycles being anovular. Lower oestrogen and progesterone levels lead to hypothalamic–pituitary hyperactivity. Disproportionately elevated FSH and frequently normal LH levels are found, their secretion remaining cyclical. Partial ovarian failure follows, as defective ovulation and corpus luteum development lead to unopposed oestrogen secretion. DUB, endometrial hyperplasia or carcinoma may follow. The next stage is either abrupt or progressive complete ovarian failure. Decreased oestrogen secretion leads to cessation of menstruation. Postmenopausally, low oestradiol (E2) but normal or elevated oestrone (E1), androstenedione and testosterone levels are found, falling gradually after 5–10 years. Secondary to increased GnRH secretion there is an increase in FSH ($\times 13$) and LH ($\times 3$), maximal 2–3 years after menses cease and declining thereafter. Decreased levels of inhibin (produced by the ovarian follicle with a specific negative feedback effect on FSH) is responsible for this disproportionate rise in FSH levels in the postmenopause.

Postmenopausal Endocrinology

E1 is quantitatively the major oestrogen in the postmenopause (ratio of E2 to E1 of 0.4), but E1 has 10% of the biological activity of E2. E2 levels correlate well with indices of vaginal maturation and superficial dyspareunia. Once the ovaries fail, the major source of oestrogen (98%) is peripheral (mainly adipose tissue) aromatisation of androstenedione from the adrenal (70%) and the ovary (30%); the remainder comes from testosterone and E2. The efficiency of conversion is two to four times greater in the 50 year old than the 20 year old. E1 levels correlate well with body weight and excess body fat. Ovarian androgen secretion increases postmenopausally; in combination with low oestrogen levels it causes postmenopausal masculinisation and hirsutism. This is more marked in obese women. Following bilateral oophorectomy (either pre- or postmenopausally) there is a 50% drop in androstenedione and testosterone levels. Lower postmenopausal levels of progesterone and 17α-hydroxyprogesterone are of adrenal origin. They bear no relationship to vasomotor symptoms. Low oestrogen levels cause prolactin levels to fall.

Anatomical Changes in the Climacteric

The shrunken, white, postmenopausal ovary has a thin cortex but relatively thick medulla with many stromal cells; these are the site of steroid hormone synthesis. The uterine body to cervix ratio returns to 1:2 as in childhood. Cystic glands may persist in the atrophic endometrium; proliferative or even hyperplastic endometrium is found in a few women with efficient peripheral conversion of androgens into oestrogens. Between 10% and 20% of women in the immediate postmenopause have genital tract evidence of oestrogen deficiency, increasing to 50–60% in women over the age of 75 years. This includes a thin, narrow vagina; shiny, atrophic vulval skin; gaping introitus; vaginal and uterine prolapse. Vaginal lactobacilli are no longer found, the pH becoming alkaline. Similar changes occur in the transitional epithelium of the bladder and urethra, which become susceptible to irritability, infection and incontinence.

Pathology of the Climacteric

Osteoporosis

Apart from oestrogen deficiency in the postmenopause, the other causes of osteoporosis include deficiency of calcium and vitamin D (dietary or malabsorption), immobilisation, excess adrenocortical secretion, drugs (e.g. steroids, cytotoxics, heparin), hyperthyroidism, hyperparathyroidism, chronic renal disease and rheumatoid arthritis.

Women over the age of 65 have a 10-fold increase in hip, spine and wrist fractures when compared with men of the same age. After the menopause there is a rise in serum calcium, phosphorus and alkaline phosphatase as well as in urinary calcium, phosphorus and hydroxyproline. This net negative calcium and phosphorus balance is reflected in the loss of 1% of the skeletal mass each year after the menopause. This can be reversed by small doses of oestrogen, which inhibit osteoclastic activity and hence bone resorption. Postmenopausal osteoporosis is particularly marked in women with thin skins. Black women are less susceptible to bone loss than white women. Osteoporotic women have significantly lower E1 and androstenedione levels. One per cent of women over the age of 65 will sustain a hip fracture, of whom one in three will die within 6 months;

10% of women aged 70 or over will have had a Colle's fracture during their postmenopausal years; 25% of white women over 60 will develop spinal compression fractures.

The routine use of dual energy X-ray absorptiometry (DEXA) measurements or other techniques to perform bone density measurements at the hip or spine cannot be justified at present. This is restricted to women with osteoporotic fractures, specific risk factors or who request hormone replacement therapy (HRT) solely for prevention or treatment of osteoporosis.

Postmenopausal bone loss can be prevented by an E/P combination. Regular exercise and calcium supplementation are thought by some to be of value, albeit less than that of steroids. E should be started within 3 years of the menopause in order to restore lost bone mass. Accelerated bone loss may occur when oestrogen is stopped. A minimum of 2 mg E2 or 20 μg ethinyl E2 is necessary to reduce the risk of fracture (risk ratio of 0.2:0.4). Prophylactic HRT for all women is advocated by some. Fast bone losers are at particular risk of osteoporosis and may be identified by measuring 24 h calcium excretion or urinary calcium:creatinine ratio. Contrary to the situation in the reproductive years, oestrogen lowers blood pressure in postmenopausal women, which is therefore not contraindicated in hypertensive women. Replacement doses of oestrogen have no adverse effect on blood coagulation nor do they increase the incidence of thromboembolic disease in postmenopausal women.

Lipoproteins and Vascular Disease

The increase in cardiovascular disease and changes in lipoprotein patterns characteristic of the postmenopause can be reversed by HRT. Doses of 2 mg of E2 valerate or 0.625 mg of conjugated oestrogens are effective in reducing deaths from ischaemic heart disease (overall risk ratio of 0.4) by preventing atherosclerosis. This should be combined with the minimum dose of P, preferably a 17α-hydroxyprogesterone derivative (pregnane type) to ensure withdrawal bleeding, minimise the risk of endometrial hyperplasia and carcinoma and avoid negating the beneficial effects of oestrogen on serum lipids and lipoproteins.

Clinical Features

The diagnosis of the menopause is usually clear on history alone. Blood FSH and E2 estimations are not required for routine clinical diagnosis but may be helpful in women who have had a hysterectomy, or who have suspected premature menopause, or to identify climacteric-related depression. Oestrogen-reversible climacteric effects include vasomotor symptoms (hot flushes, night sweats), genital and urinary tract dysfunction (vaginal dryness, dyspareunia, atrophic vaginitis, stress incontinence and the urethral syndrome (urgency, dysuria and recurrent UTIs) as well as psychological disorders. Hot flushes typically last a few minutes and are associated with sweating, palpitations and tachycardia; they may also occur at night. The aura and subjective sensation precede the mean increase in skin temperature of 2–3 deg.C. Elevated ambient temperatures cause prolonged hot flushes (up to 1 h). Flushes result from resetting of the thermoregulatory centre (thought by some to be related to activity of the GnRH centre, which controls pulsatile LH release) with adrenergic vasodilatation (plasma adrenaline increasing by 150% and noradrenaline decreasing by 40%). This leads to loss of body heat and lowering of body core temperature. Women with hot flushes (particularly if thin) have significantly lower bound and free E1 and E2 levels. The addition of clonidine (a central alpha adrenergic stimulant) or even a beta blocker may relieve hot flushes in those 5-10% of patients who do not respond to oestrogen alone.

Types of HRT and its Side-Effects

All types of oestrogen are equally effective (in equipotent doses) in relieving symptoms and have similar side-effects, but the trend is away from unconjugated synthetic oestrogen (e.g. ethinyloestradiol, which causes progressive build up of its metabolite E1 sulphate) towards natural conjugated equine oestrogen. Non-oral or implant users lose the beneficial effect on lipid and lipoprotein levels. Implants lead to high peak blood levels and the dosage cannot be adjusted. Vaginal or percutaneous creams and pessaries are well absorbed. Since they by-pass the liver they yield higher E2:E1 ratios than oral oestrogen.

Relative contraindications to HRT include previous oestrogen-related thromboembolic episodes, chronic liver disease, porphyria and hormone-dependent carcinoma, but not diabetes or hypertension.

Unopposed oestrogen users have a mean relative risk of carcinoma of the endometrium of 2%, in proportion to the duration of use (minimum latency 3-6 years) and dosage. They also have a small increase in the risk of breast cancer (latency greater than 10 years). Cyclical progesterones should be included in HRT for all women who retain their uterus. The addition of progesterone for 10-14 days per cycle reduces the risk of both carcinomas to one quarter of the natural incidence. The pregnane and gonane derivatives of 19-nortestosterone (e.g. norethisterone and norgestrel), while producing beneficial changes in lipid and lipoprotein levels, have some androgenic effects.

Postmenopausal Bleeding

Any bleeding, however small, necessitates immediate investigation by transvaginal ultrasound, endometrial biopsy or D&C.

In most cases no serious organic pathology (e.g. carcinoma of the cervix or endometrium) is found. Atrophic vulvovaginitis is common and responds to local oestrogen therapy, or a few oral cycles of combined E/P to restore vaginal acidity. Bleeding and discharge from secondary infection will cease.

Chapter 22

Virilism and Hirsutism

Virilism

Virilism includes hirsuitism, breast atrophy, male-type baldness, deepening of the voice and/or clitoral hypertrophy. Its management is described in the following section.

Hirsutism

This is abnormally located excessive growth of hair. Hyperandrogenism of adrenal or ovarian origin is the cause. Hyperprolactinaemia may cause hirsutism by the direct action of prolactin on the adrenal. Normal ovulatory menses or recent pregnancy do not exclude organic pathology. Other major causes are familial, iatrogenic (testosterone, danazol, dilantin, phenothiazines) or idiopathic. Excess response of the hair follicles to normal androgen level is generally found in the latter, but some have elevated free plasma testosterone levels of ovarian or adrenal origin. Rarely acromegaly may present with hirsutism. Even rarer are congenital anomalies such as hypertrichosis lanuginosa and the Cornelia de Lange syndrome.

Ovarian Causes of Hirsutism

Ultrasound examination demonstrates polycystic ovaries in over 90% of hirsute women even when menstrual cycles and gonadotrophin concentrations are normal. Secondary adrenal involvement explains the elevated dehydroepiandosterone often found. Normal 17α-progesterone, slightly elevated testosterone, and elevated LH and LH/FSH ratio are characteristic.

Virilising ovarian tumour may include arrhenoblastoma, thecoma, luteoma, hilus cell, granulosal or adrenal rest tumour; rarely other types are found. Though sometimes palpable, ultrasonography is generally necessary to confirm ovarian enlargement. Elevated total plasma testosterone (both free and bound to sex hormone binding globulin (SHBG)), in combination with normal 17α-progesterone and DHEAS are typical.

Adrenal Causes of Hirsutism

These include tumours, hyperplasia and Cushing's syndrome.

The origin (zona glomerulosa, fasciculata, reticularis) determines the clinical presentation of adrenal tumours. Tumours involving more than one area result in a mixed clinical picture that may include primary aldosteronism (hypertension and hypocalaemia) and Cushing's syndrome. Substantially elevated plasma 17α-progesterone and DHEAS and, to a lesser extent than in ovarian pathology, testosterone, point to an adrenal cause. A fall in plasma 17α-progesterone and urinary 17α-oxosteroids on administering dexamethasone indicates that adrenal hyperplasia is the underlying cause. Partial suppression may occur in cortical tumours. Increased urinary free cortisol, together with loss of diurnal rhythm and inability to suppress the morning peak of cortisol with 2 mg of dexamethasone confirms the diagnosis of Cushing's syndrome.

The cause of excess cortisol secretion must be determined: hypothalamus, pituitary or adrenal. Skull and pituitary fossa radiographs, whole body CT and sometimes ultrasound may be of value.

Treatment of Hirsutism

Organic pathology should be dealt with appropriately. Weight loss may help obese hirsute women; mechanisms include lowering insulin, thus increasing SHBG (and thereby reducing free androgen levels) and decreasing ovarian thecal cell response to LH. Ethinyl oestradiol (35–50 μg), in combination with norethisterone, may improve hirsutism by suppressing ovarian androgen production and increasing SHBG. By depressing SHBG, norgestrel-containing oral contraceptives have the opposite effect.

Cyproterone acetate (50–150 mg on days 5–15 of the cycle), usually prescribed with ethinyl oestradiol (50 μg on days 5–26), competitively inhibits testosterone binding to the cytoplasmic receptor, thus blocking its intracellular action. Long-term treatment (12–18 months) is required to achieve an approximately 70% cure rate. The two drugs are combined in the preparation Dianette, which is effective as maintenance treatment, but rarely sufficient to reverse excess hair growth. Spironolactone (100 mg from day 5 to day 28) also acts as a competitive androgen inhibitor but in addition it inhibits cytochrome P450, essential for androgen biosynthesis. It is not suitable for long-term use. Both drugs interfere with sexual differentiation of a male fetus. Cyproterone acetate may exacerbate hyperinsulinism and alter lipid profiles, characteristics of lean women with polycystic ovaries. Hirsute patients are thus at increased risk of cardiovascular disease. Newer non-steroidal anti-androgens (e.g. flutamide) are undergoing evaluation. Most women benefit from symptomatic treatment with depilation and cosmetics.

Chapter 23

Genital Infections

Infections and Related Conditions of the Vulva

Pruritus Vulvae

Pruritus of the vulva may be due to: (a) vaginal infection (*Candida* and *Trichomonas*); (b) chemical irritants (soaps, antiseptics, detergents); (c) diabetic vulvitis; (d) dystrophies and malignant lesions; (e) ulcerative conditions (see below); (f) psychological causes; (g) condyloma acuminata; (h) systemic disease (obstructive jaundice, thyroid disease, Crohn's disease, leukaemias); (i) generalised skin disease (eczema, urticaria, scabies); and (j) atrophic vaginitis (common in postmenopausal and breast feeding women).

Treatment depends on the cause. Symptomatic management includes application of zinc and castor oil ointment, or 1% hydrocortisone cream or ointment.

Ulceration of the Vulva

Granuloma Inguinale

This is due to venereal infection with the bacterium *Donovania granulomatosis* (Donovan bodies in mononuclear cells) and occurs in black women in Africa, the Caribbean and southern USA. The primary lesion is an itchy papule that ulcerates. Lymph gland enlargement is not a prominent feature. Treatment is with tetracycline or ampicillin (500 mg, four times daily for three weeks).

Lymphogranuloma Venereum

This is a venereal disorder due to L 1-3 serotypes of *Chlamydia trachomatis*, common in tropical countries, and commoner in men than in women. Swelling of the inguinal lymph glands follows a few weeks after the primary vulval ulcer. Complications are proctitis, elephantiasis and carcinoma. Diagnosis is by a complement fixation test or a microimmunofluorescent test on serum. Treatment is with tetracyclines or erythromycin.

Behçet's syndrome

This is a connective tissue disorder with multisystem involvement. These include aphthous ulcers of the mouth and vulva. It is sometimes associated with conjunctivitis and panophthalmitis. Treatment is with oral or topical corticosteroids.

Other conditions associated with oral and vulval ulceration include ocular pemphigus (in older women) and erythema multiforme (with bullous lesions on extensor surfaces and palms and soles).

Herpes Simplex Virus (HSV) Infection

This is due to type II virus (HSV2) (or occasionally HSV1, which more commonly causes lip lesions) acquired by sexual contact. Asymptomatic shedding of the virus may occur, but rarely during recurrences. Lesions are commonest on the mucosal surface of the labia minora but often spread to surrounding skin; a milky vaginal discharge may be present. The cervix is infected in 70% of cases but often appears normal. Perianal lesions and proctitis may also occur. In a primary attack the characteristic lesion is an eruption of painful vesicles that soon break down to form ulcers; healing occurs in 2–6 weeks. There is general malaise and lymphadenopathy. Fresh crops of vesicles may occur over 2–4 weeks. Patients are infectious when lesions are present. Sacral radiculitis is a rare complication. Dysuria is a prominent feature and may even lead to urinary retention. The condition recurs in 30–70% of cases due to activation of latent infection by factors such as trauma, stress and ultraviolet light. Secondary attacks are milder than primary and heal quicker. The diagnosis can be confirmed using scrapings from fresh lesions (vulva and cervix), and immunofluorescence or cytopathic effects on tissue culture. Patients also develop antibodies within 7 days of infection. The treatment is with oral acyclovir (200 mg five times daily for 5 days). This does not prevent recurrence. One in 100 000 women are HSV infected at term: 50% of their infants will become infected during vaginal delivery. Delivery is by caesarean section either electively or up to 4 h after ruptured membranes for: (a) active lesions, either primary or recurrent; or (b) positive viral cultures without lesions.

Rare Causes of Vulval Ulceration

These include trauma (scratching), tuberculosis, chancroid (*Haemophilus ducreyi*) and syphilis (*Treponema pallidum*).

Syphilis is the commonest cause of painless genital ulcers: most cases occur in homosexuals. The primary lesions (chancre) appear 10–90 days after sexual contact. They may be: (a) genital (labia, fourchette, clitoris, cervix) or extragenital (painless inguinal lymphadenopathy). Secondary syphilitic lesions occur 4–8 weeks later. These may be:

1. Symmetrical, non-itchy maculopapular rash.
2. Condylomata lata (large, fleshy masses on anus, labia).
3. Shallow painless ulcers on mucous membranes (snail-track).
4. Constitutional symptoms (malaise, anorexia, fever, lymphadenopathy).
5. Rarely: hepatitis, iritis, neurological disease.

If untreated, cardiovascular (10%), neurological (15%) and gummatous (15%) lesions will appear after a latency of 2–20 years; 65% of untreated cases have no clinical sequelae. Diagnosis of primary syphilis is by identification of the spirochaete by darkground microscopy; serology may be negative for 3–5 weeks after infection. Serological tests may be non-specific (VDRL, largely replacing the Wasserman reaction) or specific (treponemal agglutination (TPHA); fluorescent treponemal antibody (FTA-ABS)). Biological non-specific false positive reactions occur after acute (e.g. measles, mumps, chicken pox) and chronic (leprosy) infections, or after immunisation (typhoid, yellow fever), or in autoimmune disease, or in other treponemal conditions (yaws, bejel, pinta). The specific tests may be negative in primary syphilis and neither can distinguish between the different treponemal conditions; all serological tests may remain positive after adequate treatment.

Prevention and control is by (a) contact tracing, (b) screening donated blood, and (c) testing patients in antenatal and genitourinary clinics. Treatment is with procaine penicillin (600 000 units/day for 10 days). If the woman is allergic to penicillin, erythromycin or tetracycline are used and repeated after 3 months. Fetal infection (congenital syphilis) is rare and may occur at any time during pregnancy.

Genital Warts

Genital warts (condyloma acuminata) are caused by human papilloma virus (HPV) types 6 and 11. They are most commonly found in the fourchette or perianal area. Lesions can regress but usually persist or enlarge, e.g. during pregnancy, for many years. Coexisting vaginal and/or cervical warts can be identified colposcopically in two-thirds of women with vulval warts. Treatment is with application of small quantities of 15–20% podophyllin (except during pregnancy), or 40–100% trichloroacetic acid (TCA); surgical or laser excision; cryocautery; or interferon. TCA may be used during pregnancy.

Vaginal Infections

Genital symptoms and signs are unreliable diagnostic features. Their use alone would miss 90–95% of cases of candidiasis and trichomoniasis. Microbiological examination is essential.

Trichomonas Vaginitis

Trichomonas vaginalis is a one-celled organism with four anterior flagella. It is transmitted by sexual contact and causes one third of all cases of vulvovaginitis. The discharge is classically described as profuse, greenish, frothy, malodorous (fishy); the vaginal walls may show small "strawberry" spots, and on colposcopy typical "Y-shaped" vessels are seen. Contact bleeding is typical. The presence of the organism is confirmed by simple microscopy. Treatment (of both partners) is with oral metronidazole (200 mg three times daily for 7 days, or a single 2 g dose). This can cause nausea and dizziness, and in some cases a disulfiram-like reaction with alcohol. It should not be used in the first trimester of pregnancy.

Candida Vaginitis

Candida albicans is a Gram-positive yeast that forms mycelia and spores and is a normal commensal of the mouth, gastrointestinal tract and the vagina. Infection results from overgrowth, which is commoner in the presence of pregnancy, diabetes, and prolonged treatment with corticosteroids, immunosuppressive agents and antibiotics. The incidence has greatly increased in recent years. The discharge is thick, white and curdy; there may be intense inflammation and pruritus, dysuria and dyspareunia. The clinical symptoms appear to be a reaction to an allergen or endotoxin produced by the yeast. The organism is easily seen in a wet smear after destruction of other cells with 10% potassium hydroxide (KOH). It can be cultured on glucose agar from swabs transported in Stuart's or Sabouraud's medium. The treatment is with imidazole pessaries (clotrimazole or miconazole) for 3–7 days. Ketoconazole provides effective oral therapy. Both drugs should be avoided during pregnancy. Natural remedies include yoghurt and garlic. Failure of treatment is due to either inadequate therapy or reinfection from male partner. Both partners should therefore be treated.

Gardnerella Vaginitis

Gardnerella (formerly *Haemophilus vaginalis*) is a Gram-negative facultative anaerobic bacillus that may cause a frothy, grey, malodorous discharge (the odour increases if the smear is treated with 10% KOH). Epithelial cells with bacilli attached to the surface ("clue" cells) are characteristic. It can be cultured on Casman's agar from swabs transported on Casman's broth. The treatment is the same as for *Trichomonas*.

Other Types of Vaginitis

During reproductive years the vaginal pH is acid due to conversion of glycogen into lactic acid by Doderlein's lactobacillus (follicular phase pH 3.2–4.2; luteal phase, pH 5.5; menstrual pH 6.5–7.5). Thus pyogenic infections are rare at this time, except in the presence of a foreign body or following overzealous douching. Pyogenic infections are not uncommon in childhood or in association with atrophic vaginitis in the elderly, and are treated with topical oestrogens. In children, vaginal discharge may occur in association with threadworm infestation (treated with piperazines).

Gonococcal vaginitis is caused by the Gram-negative diplococcus *Neisseria gonorrhoea*. The risk of transmission is threefold higher in blacks than in whites. The latent period varies from a few days to 8 weeks. Most women are asymptomatic, but there may be pain, inflammation and a purulent discharge from the urethra and cervix. Bartholinitis may occur. A child delivered by a woman with active gonorrhoea is at risk of ophthalmia neonatorum. The diagnosis is made by Gram-staining of swabs from the cervix and urethra (intra- and extracellular diplococci are seen); swabs can be transported in Stuart's transport medium and cultured on Thayer–Martin plates or on modified New York City medium.

Treatment is with penicillin (2.4 mu of procaine penicillin into each buttock; 1 g of probenecid by mouth) or tetracycline (250 mg four times a day for 10 days) or spectinomycin (4 g) in penicillin-sensitive patients, or those who have a β-lactamase-producing strain.

Cervicitis

The red, vascular columnar epithelium of the endocervix is commonly visible on speculum examination of women during reproductive years, and various amounts of cervical mucus may be present. This normal appearance is often given pathological names such as "erosion" and "ectropion". Excess mucus production, especially if it appears purulent, is referred to as cervicitis. Small cysts (Nabothian follicles) may also be apparent. Any unusual appearance should be investigated for the presence of cancer. If the lesion is benign and thought to be responsible for the vaginal discharge it may be treated with cautery or cryosurgery, the latter having the advantage of little discomfort and rapid healing.

Cervical warts are due to HPV types 6 and 11.

Pelvic Infections

Acute Pelvic Infections

These affect 1-2% of women aged 15-39 years in Western industrialised countries. These are most commonly due to ascending sexually transmitted diseases (STDs) such as gonorrhoea (20%) and chlamydia (50%) (see below), especially at the time of menstruation. They may also occur due to retention of necrotic tissues following abortion or delivery, or following iatrogenic trauma (e.g. D&C, salpingogram, insertion of IUCD). The incidence is lower in women on oral contraceptive agents. The commonest non-STD organisms are Gram-negative bacilli, streptococci, staphylococci, and anaerobes such as *Clostridium* and *Bacteroides*. There is rapid spread from the uterus (endometritis) to adjacent tissues (interstitial salpingitis, oophoritis, thrombophlebitis in the parametrium, and pelvic peritonitis). The risk of tubal occlusion is 13% after one attack, rising to 75% after three or more attacks.

The clinical features include those of general infection. There is pelvic pain and tenderness, especially on movement of the cervix; the parametrium is thickened, the uterus can be enlarged and tender, and there may be signs of pelvic peritonitis. There is often an offensive brownish vaginal discharge. A full set of swabs should be taken for microbiological analysis (vaginal, urethral and cervical). The diagnosis can usually be confirmed or excluded by laparoscopy. Pelvic abscesses can be identified by ultrasound.

The treatment is rehydration and antibiotics (ampicillin 500 mg 6-hourly for 2 weeks; metronidazole 400 mg 8-hourly for 14 days; and sometimes gentamicin). This treatment may be varied according to the results of culture, if and when available, or if the patient does not respond. If *Chlamydia* is thought to be involved, a tetracycline (doxycycline) or erythromycin should be added. The latter can be given intravenously. Both partners should be treated.

Gonococcal Infection

Upward spread of gonococcal infection of the lower genital tract is common and usually asymptomatic. The epithelium of the fallopian tube is usually involved, with subsequent occlusion. The clinical features are those of primary gonococcal infection together with those of an acute pelvic infection (see above). Gonococcal salpingitis is commonest in the first half of the cycle.

Chlamydial Infections

Chlamydia trachomatis (an obligate intracellular parasite; serotypes D–K) is the commonest sexually transmitted organism in the developed world. *Chlamydia* contain both DNA and RNA and replicate by binary fission using the host cell energy system. Evidence of infection is found in 20–25% of women attending STD clinics, including 30–50% of women with gonorrhoea. Many cases are silent but it may cause urethritis in both men and women (50% of cases of non-gonococcal urethritis), epididymitis, bartholinitis, cervicitis, a fibrous perihepatitis (Fitz-Hugh–Curtis syndrome), meningoencephalitis, inclusion conjunctivitis in the newborn, proctitis and salpingitis (40% or more of cases of acute pelvic inflammatory disease (PID)). The infection can be diagnosed in a cervical smear by immunofluorescence. Both partners should be treated with either tetracycline or erythromycin (during pregnancy) (250 g four times daily for 10 days). Between 2% and 12.5% of pregnant women have chlamydia; 30–50% of their children develop conjunctivitis and 10–20% pneumonia.

Other Infections

Acute PID can result from ascending infections by *Mycoplasma hominis* (10%), *Ureaplasma urealyticum* (5%) and anaerobes (5%). Treatment is with tetracycline or erythromycin as for *Chlamydia*.

Chronic Pelvic Inflammatory Disease

Unless treated rapidly and vigorously, acute pelvic infections can become chronic. There are pelvic adhesions and occlusion of the fallopian tubes. The latter may form a pyosalpinx or a sterile hydrosalpinx; further extension may lead to a tubo-ovarian abscess.

The main clinical features are pain (sometimes associated with menstruation or intercourse), menstrual irregularities (polymenorrhagia) and infertility. In long-standing cases there is weight loss. On examination there is thickening of the parametrium and tender pelvic masses may be found. The chronic course may be punctuated by acute exacerbations. The condition is much overdiagnosed. Treatment is one or more courses of antibiotic therapy, identical with that for an acute infection (see above). Surgery may be indicated if the response to medical treatment is inadequate and may include a variety of procedures from simple salpingectomy and division of adhesions to total pelvic clearance. If infertility is the main problem then various forms of tubal surgery are indicated. Assisted conception techniques may be necessary.

Rupture of a tubo-ovarian abscess presents as an acute emergency, which is managed with immediate surgical drainage under antibiotic cover.

Genital Tuberculosis

Genital tuberculosis usually occurs as a result of haematogenous spread from a pulmonary focus. The commonest lesion is an interstitial infection of the medial portion of the tube; the tube is usually patent but narrow and rigid. The endometrium may also be involved (90% of cases); lesions in other sites are rare, but a "plastic peritonitis" has been described.

The clinical features include those of chronic pelvic inflammatory disease. Very rarely, endometrial involvement may lead to amenorrhoea. The treatment is isoniazid, rifampicin, pyrazinamide and ethambutol or streptomycin for 8 weeks, followed by isoniazid and one of the other drugs for 6–18 months.

Other Sexually Transmitted Diseases (Non-genital)

HIV Infection

HIV is transmitted through: (a) sexual activities (male to female and vice versa); (b) blood products; (c) perinatally (in utero, intrapartum, post partum); and (d) breast milk. Some patients have symptoms of acute infection (fever, malaise, myalgia, lymphadenopathy, pharyngitis, rash); others do not. Chronic infection is largely asymptomatic but intermittent constitutional symptoms may occur (fever, night sweats, diarrhoea, weight loss). After a latency of 2–5 years 6–30% of chronic HIV-infected patients progress to AIDS. These present with (a) tumours (Kaposi's sarcoma, non-Hodgkin's lymphoma, mouth and anorectal squamous carcinoma) or (b) opportunistic infections with viruses, bacteria, fungi and protozoa.

Depletion or impaired function of lymphocytes bearing T4 or cluster differentiation (CD) antigen 4 (subsets of the T helper–inducer lymphocyte) is the central immunological abnormality. HIV is a retrovirus that can enter only cells bearing the CD4 antigen. The virus makes a DNA copy of the RNA genome (proviral DNA), which integrates into host cell DNA.

Prevention and control of infection requires (a) surveillance, (b) health education, (c) screening, and (d) counselling. The onset and severity of clinical AIDS may be delayed and reduced by combination antiretroviral treatment, namely nucleoside reverse transcriptase inhibitors (zidovudine), non-nucleoside reverse transcriptase inhibitors (nevirapine) and protease inhibitors (indinavir).

Viral Hepatitis

There are at least five types of "infectious" hepatitis: A, B, C, D and E. In addition other viral agents such as infectious mononucleosis and cytomegalovirus can cause hepatitis as part of a more generalised infection. Infective hepatitis (virus A (picornavirus), incubation 15–40 days) occurs epidemically in institutions and in developing countries; it is excreted in faeces (in which it can be detected by an immunoassay) and spread by contaminated food and water. There is no carrier state. Hepatitis B (virus B (DNA) incubation 60–160 days) is usually transmitted by inoculation of blood products (e.g. transfusion, contaminated needles, drug addicts or by close person-to-person contact such as mother and baby). Persons handling blood or working in haemodialysis units are therefore at special risk. Both hepatitis A and B may be sexually transmitted.

Three antigens are associated with hepatitis B (HB) infection.

1. *HBsAg*: the outer coat of the virus (visualised by electron microscopy as the Dane particle) has a surface antigen (HBsAg: originally known as Australia antigen) that is always present in the acute phase of infection and also in the chronic carrier. The appearance of antibody to HbsAg indicates recovery and subsequent immunity.
2. *HBcAg*: the core antigen which is not detected in serum. IgM antibodies to HBcAg appear during acute infection and IgG antibodies indicate past exposure.
3. *HBeAg*: the antigen associated with infectivity. It is detectable in serum during acute infection and in carriers with a high risk of transmission. Antibodies to HBeAg indicate a low risk of transmission.

Many patients who have been infected with hepatitis B virus, either overtly or subclinically, continue to excrete the virus (carrier rates range from 0.1% in the UK to 15% in the Far East and Africa). HBsAg carriers who continue to express e antigen are considered to be more infectious than others, and to have a higher risk of chronic liver disease and hepatoma. HBsAg-positive mothers with the e antigen or lacking anti-e antibodies may infect infants after delivery. In acute cases the virus can cross the placenta. Thus acute infection with hepatitis B may be transmitted to the fetus, especially in the third trimester, and cause stillbirth or neonatal hepatitis. Passive immunisation with immunoglobulins containing anti-HBs antibodies is indicated for babies of e-antigen-positive mothers and those with HBsAg, but no anti-e antibodies. This will prevent these infants from becoming carriers. Passive immunisation should be followed by active immunisation with hepatitis B vaccine.

Chapter 24
Benign and Malignant Lesions of the Vulva

Benign Tumours of the Vulva

A variety of benign epidermal lesions may be encountered, including epidermal and sebaceous cysts, seborrhoeic keratoses, squamous papillomata and pigmented naevi.

Vulval warts occur in women of reproductive age, and are found mainly on the labia majora, perineum, and perianal skin. They are venereal in origin and usually associated with other vulvovaginal infections. The larger lesions are referred to as condyloma acuminata. Histologically the epithelium is thickened and folded around connective tissue stalks. Treatment is by application of 20% podophyllin or diathermy.

Unusual benign tumours of the vulva include hydradenoma (from apocrine sweat glands), fibroma, lipoma, neurofibroma, leiomyoma, granular-cell myoblastoma (from nerve sheaths, also found in the tongue), haemangioma and lymphangioma.

Other Benign Vulval Swellings

Various benign cysts may be found in the vulva (Table 24.1). Bartholin's glands lie in the posterior third of the labium majus and the collecting ducts open distal to the hymen at 5 and 7 o'clock on the circumference. If the duct is obstructed as a result of trauma or infection, a cyst forms that is lined by transitional epithelium.

Table 24.1. Benign cysts

Type	Features
Wolffian (Gartner's) duct	Around hymen or clitoris
Canal of Nuck	Upper part of labium majus
Epidermal inclusions: Skene's duct Bartholin's	See text

An abscess with pyogenic organisms may form in association with gonorrhoea. Treatment is by drainage and marsupialisation. Recurrence may be treated by removal of the glands.

Non-cystic swellings include haematomas and urethral caruncles. Caruncles may be difficult to distinguish, if they can be distinguished at all, from urethral prolapse. Three types are described – angiomatous, polypoid and granulomatous – the last being associated with infections. They cause pain and occasional blood-staining and may be exquisitely tender to touch. Treatment is excision by diathermy.

Vulval Dystrophies

The terms used to describe these conditions and their commonly accepted meanings are given in Table 24.2. General features of these conditions are as follows. They occur predominantly after the menopause. They are associated with white patches and fissures on vulval skin. They affect the labia, clitoris and perianal region but not the vagina and they may cause severe pruritus.

The histological features of hypertrophic dystrophies in the epidermis are: (a) hyperkeratosis; (b) thickening; (c) enlargement, down-growth and forking of rete pegs; and (d) thickening of the malpighian layer. In the dermis the features are (a) disappearance of elastic fibres, (b) round cell filtration, and (c) oedema and hyaline replacement of collagen tissue. The clinical appearance of the vulval conditions is often very similar and the distinction can be made only by multiple-site biopsy. Review in a specialist clinic with input from dermatologists as well as gynaecologists is important.

Conditions that may be associated with vulval dystrophies include psoriasis, fungal infections, dietary deficiencies (iron, folate, riboflavin, vitamin B12), allergies and diabetes. Usually no underlying cause is found.

Treatment of atrophic dystrophies includes general measures (hygiene, general health, diet, etc.) and locally applied steroid creams of varying strengths. If the hyperplastic elements are thought histologically to show significant evidence of premalignant change then the affected area should be excised.

Premalignant Conditions of the Vulva

Hyperplastic vulval dystrophies may show histological evidence of atypia up to and including carcinoma-in-situ (numerous mitotic figures, loss of stratification,

Table 24.2. Meaning of various terms used in the description of vulval dystrophy; mixed forms may also occur

Term	Meaning
Hypertrophic dystrophy	See text
Atrophic dystrophy	As hypertrophic but with atrophy of epithelium and flattening of rete pegs and malpighian layer
Lichen sclerosus	As atrophic dystrophy, but may appear on multiple areas of the body and can occur at all ages including childhood
Vaginal intrapithelial neoplasia	Dyplastic keratinocytes confined to the epithelium

cellular pleomorphism) without invasion through the basement membrane. Variants of this include Bowen's disease and Paget's disease, in which a localised lesion may persist without spreading for many years. The lesions of Paget's disease contain characteristic large, pale Paget's cells, an appearance comparable to that of Paget's disease in the nipple. The treatment of carcinoma-in-situ is wide excision. In younger patients a more limited operation is possible, with frozen sections at the time of operation to ensure that excision is complete; the defect may be filled with a skin graft, but primary closure is usually possible. If pruritus is present preoperatively it always recurs within 2 years. Up to 25% of Paget's are associated with a second pelvic carcinoma, usually of the rectum.

Carcinoma of the Vulva

This represents 3-5% of malignancies of the female genital tract; some 15-30% of patients have had or will develop cancer of the cervix. Most cases occur in women over 50; the mean age is 60. It may develop from an area of carcinoma-in-situ or hyperplastic dystrophy. The growth is a squamous cell carcinoma that is found most commonly in the anterior half of the labia majora. Metastasis occurs via lymphatics to superficial and deep layers of lymph nodes. The superficial nodes include the medial and lateral inguinal and the medial and lateral femoral; the deep nodes include the inguinal, femoral, obturator and external iliac. The femoral nodes (node of Cloquet) form a central drainage point for most of the lymphatics of the vulva. Spread may be ipsilateral or bilateral. Some 15-30% of patients with no clinical evidence of lymphatic spread have metastatic cancer in the lymph glands. Moderately enlarged nodes may be the result of local infection rather than malignant spread. Direct local invasion may involve any of the surrounding organs. Staging is by the International Federation of Gynaecologists and Obstetricians (FIGO) system (Table 24.3).

The characteristic symptoms are pruritus and pain, less commonly a lump, bleeding or discharge. On examination the lesion may be hypertrophic, ulcerative, nodular, or some combination of these.

Treatment is by wide excision or vulvectomy with bilateral inguinofemoral lymphadenectomy; selected cases may also have pelvic lymphadenectomy. Primary closure can usually be achieved but, failing this, skin grafting may be employed. In cases involving the anal margin posterior exenteration and

Table 24.3. Staging of vulval cancer (FIGO 1988)

Stage 0	Carcinoma in situ, intraepithelial carcinoma
Stage IA	Tumour confined to the vulva and/or perineum < 1 mm invasion; < 2 cm in greatest dimensions, nodes are not palpable
Stage IB	As IA but invasion > 1 mm
Stage II	Tumour confined to the vulva and/or perineum, > 2 cm in greatest dimensions, nodes are not palpable
Stage III	Tumour of any size with: (a) adjacent spread to the lower urethra and/or the vagina, or the anus; and/or (b) unilateral regional lymph node metastasis
Stage IVA	Tumours invade any of the following: upper urethra, bladder mucosa, rectal mucosa, pelvic bone and/or bilateral regional node metastasis
Stage IVB	Any distant metastasis including pelvic lymph nodes

colostomy may be required. Some 80-90% of cases are suitable for major surgery. There is an operative mortality of 2-5% due to pulmonary embolism, infection and haemorrhage. Less serious complications include lymphoedema (lymphocysts), and inguinal/femoral hernias. Five year survival rates are 50-60% if lymph nodes are involved and 70-80% if they are not. Radiotherapy may be used to reduce the size of the primary lesion or to treat recurrence; however, vulval skin is particularly sensitive to radiation and surgery is the main treatment, as with other squamous cell skin cancers.

Rare Malignant Tumours of the Vulva

These include:

1. Basal cell carcinoma (2-4% of vulval malignancies; treated by local excision).
2. Melanoma (4%; treated by radical vulvectomy and inguinal/femoral lymphadenectomy).
3. Sarcoma (2%; occurs in a younger age group (under 40 years) and distant recurrence is common).
4. Adenocarcinoma or squamous cell carcinoma of Bartholin's gland (treated by extensive local excision and lymphadenectomy; 5-year survival less than 10%).
5. Metastases from adjacent organs (treated by local excision).

Chapter 25
Benign and Malignant Lesions of the Vagina

Benign Tumours of the Vagina

These include papilloma, fibroma, fibromyoma and neurofibroma. The commonest are cysts of the mesonephric (Gartner's) and paramesonephric (müllerian) ducts. Paramesonephric cysts are usually single and occur in the subepithelial connective tissue of the upper vagina, with a lining similar to the cervix of the fallopian tube. Mesonephric cysts are sausage shaped and lie anterolaterally; they are lined by cubical epithelium and surrounded by smooth muscle.

Vaginal adenosis (or adenomatosis) consists of multiple hyperplastic mucus-secreting glands. There is leucorrhoea and occasional bleeding. This condition has been associated with stilboestrol treatment of the patient's mother during pregnancy, and the lesion may precede a clear cell adenocarcinoma of the vagina.

Carcinoma of the Vagina

This is rare (1-2% of genital tract malignancies) and occurs mainly in older women (mean age 65 years). Most cases are squamous cell carcinoma with a few adenocarcinomas arising from the associated ducts; the staging is shown in Table 25.1. The commonest site is the posterior wall of the upper half of the vagina. Lymphatic spread is to the deep pelvic nodes (upper two thirds) and the inguinal nodes (lower third). Involvement of bladder and rectum is a late event.

The symptoms are bleeding and discharge. Treatment of lesions in the upper third of the vagina may include: (a) radical hysterectomy and pelvic lymphadenectomy with removal of all or most of the vagina; (b) radiotherapy (intracavitary or external); or (c) a combination of (a) and (b). Involvement of adjacent organs may indicate exenteration. Tumours of the lower third of the vagina can often be dealt with by radical vulvectomy and inguinal lymphadenectomy. The 5 year survival rate is 85% for stage I decreasing to 40% for stage III and 0% for stage IV.

Carcinoma of the vagina can also be metastatic as a result of spread from adjacent organs; this is commoner than primary carcinoma.

Table 25.1. Staging of carcinoma of the vagina (FIGO 1988)

Stage 0	Carcinoma-in-situ; intraepithelial carcinoma
Stage I	Limited to the vaginal wall
Stage II	Involves the subvaginal tissues but does not extend to the pelvic wall
Stage III	Extends to the pelvic wall
Stage IVa	Tumour involves mucosa of bladder or rectum and/or extends beyond the true pelvis. NB: The presence of bullous oedema is not sufficient evidence to classify a tumour stage IV
Stage IVb	Distant metastasis
Mode of spread:	Upper two thirds of vagina → pelvic nodes Lower one third of vagina → inguinal nodes

Chapter 26
Benign and Malignant Lesions of the Cervix

Benign Tumours of the Cervix

Cervical polyps consist of mucus glands in a connective tissue stroma covered by columnar or squamous epithelium. They appear as cherry-red swellings, usually arising within the canal, and can cause discharge and bleeding. They can be removed by excision or avulsion with diathermy to the base; the operation should be performed under general anaesthesia.

Carcinoma-in-situ of the Cervix (see FIGO staging, Table 26.1)

This lesion arises at the squamocolumnar junction and is distinguished from benign hyperplastic lesions by the histological appearance including loss of stratification and cellular pleomorphism (variation in size and shape of cells; increased nuclear/cytoplasmic ratio; bizarre mitoses). Cervical intraepithelial neoplasia (CIN) is divided into three grades; CIN I, mild dysplasia; CIN II, moderate dysplasia; CIN III, severe dysplasia and carcinoma-in-situ. The commonest age group for CIN overall is 20–28 years, and for carcinoma-in-situ is 30–39 years.

The possibility of CIN is identified by cytological screening of the cervix (the final diagnosis of CIN can be made only by biopsy). Routine screening of this type should commence at 25–30 years, to be repeated 1 year later, and thereafter at intervals of 3–5 years until age 70.

The false-negative rates of an adequate test are less than 3%, but rise to 25% under many clinical circumstances. The correlation between smear results and histology is poor; about 15% classed as mild dyskariosis will be CIN III on histology. Overestimation of the stage by histology is less common. Positive smears are evaluated by colposcopy and biopsy. Appropriate areas can be selected as those which stain with acetic acid and fail to stain with Schiller's iodine. This eliminates the need for cone biopsy unless the squamocolumnar junction cannot be seen or if microinvasion cannot be excluded. The colposcopic features of microinvasive carcinoma are (a) a dense white appearance after application of acetic acid, (b) coarse punctation, and (c) a mosaic appearance. Punctation

Table 26.1. Staging of carcinoma of the cervix (FIGO 1988)

Stage	
0	Carcinoma-in-situ, intraepithelial carcinoma; cases of stage 0 should not be included in any therapeutic statistics for invasive carcinoma
I	The carcinoma is strictly confined to the cervix (extension to the corpus should be disregarded)
Ia	Preclinical carcinoma of the cervix; that is, those diagnosed only by microscopy
Ia1	Invasion < 3 mm deep, 7 mm wide
Ia2	Lesions detected microscopically that can be measured; the upper limit of the measurement should not show a depth of invasion of > 5 mm taken from the base of the epithelium, either surface or glandular, from which it originates; a second dimension, the horizontal spread, must not exceed 7 mm; larger lesions should be staged as Ib
Ib	Lesions of greater dimension than stage Ia2 whether seen clinically or not; preformed space involvement should not alter the staging but should be specifically recorded so as to determine whether it should affect treatment decisions in the future
Ib1	Lesions < 4 cm
Ib2	Lesions > 4 cm
II	The carcinoma extends beyond the cervix, but has not extended on to the pelvic wall; the carcinoma involves the vagina but not as far as the lower third
IIa	No obviously parametrial involvement
IIb	Obvious parametrial involvement
III	The carcinoma has extended onto the pelvic wall; on rectal examination there is no cancer-free space between the tumour and the pelvic wall; the tumour involves the lower third of the vagina; all cases with a hydronephrosis or non-functioning kidney should be included, unless they are known to be due to other causes
IIIa	No extension onto the pelvic wall, but involvement of the lower third of the vagina
IIIb	Extension onto the pelvic wall or hydronephrosis or non-functioning kidney
IV	The carcinoma has extended beyond the true pelvis or has clinically involved the mucosa of the bladder or rectum
IVa	Spread of growth to adjacent organs
IVb	Spread to distant organs

Staging in cervical cancer is clinical, i.e. it does not rely on surgicopathological results, as many women are treated with radiotherapy. All other gynaecological cancers are surgically staged.

consists of small dark dots that become confluent in the mosaic pattern. These changes are caused by the altered appearance of subepithelial capillaries.

The prevalence of carcinoma-in-situ in a previously unscreened population is 2–4 in 1000. The incidence of clinical invasive carcinoma has fallen pari passu with the introduction of routine screening (e.g. from 24 to 3.5 in 100 000 in British Columbia). The incidence of progression has been variously reported as 30–75% and the interval between the preinvasive and invasive lesions averages 12 years.

In the UK, the incidence of invasive cancer has diminished very little over the last decade. However, the incidence of positive smears has increased threefold.

Microinvasive Carcinoma (see FIGO staging, Table 26.1)

This is a grade of lesions lying between carcinoma-in-situ and frankly invasive tumours, in which there is minimal penetration of the epithelial lesion through the basement membrane (less than 5 mm of stromal invasion). The risk of spread beyond the cervix increases directly with the depth of the invasion, the surface area of the lesion, and the presence of lymph or blood vessel involvement. The risk of spread is less than 1%, with less than 1 mm of stroma invasion, rising to 5% at 5 mm and more if lymph channels are involved. The diagnosis of microinvasive carcinoma can only be made on the basis of cone biopsy, not colposcopically directed biopsy alone. If the diagnosis is confirmed and the resection line is clear of the abnormality then no further treatment is required other than cytological follow-up. If there is any doubt about the completeness of removal a radical hysterectomy is performed.

Carcinoma of the Cervix (see FIGO staging, Table 26.1)

This is the commonest tumour of the genital tract in most countries. It is responsible for 0.5–1% of deaths in women, though the mortality rate is falling and is less than that of ovarian cancer. The commonest age group is 35–64 years. Some 70–90% of cases are squamous cell lesions and 10–20% are adenocarcinomas; one third of the latter contain squamous elements. Adenocarcinomas usually arise in the endocervix and a subgroup, the clear cell carcinoma, is the commonest uterine tumour in patients under 21 years; presentation and treatment are the same as for squamous carcinoma. Squamous cell carcinomas can be divided into keratinising and non-keratinising types, and the latter into large- and small-cell types; the commonest form is the large-cell non-keratinising. A special type of squamous carcinoma is the verrucoid cancer, which is very slow growing and radio resistant. The prognosis is good after local resection. Factors that affect the incidence/aetiology of cervical cancer are shown in Table 26.2.

Early spread is by local infiltration. Invasion of surrounding organs is a late event and uraemia due to obstruction of the ureters is the commonest cause of death. Lymphatic spread occurs first to the parametrial, paracervical, vesico and rectovaginal, hypogastric, obturator and external iliac glands; it is found later in the sacral, common iliac, vaginal and para-aortic glands.

The commonest presenting symptom is bleeding (intermenstrual, postcoital or postmenopausal). Discharge, pain and disturbances of micturition and defaecation occur only in late stages. The appearance may vary from that of a

Table 26.2. Factors affecting the incidence/aetiology of squamous cell carcinoma of the cervix

Factor	Effects
Parity	Commoner in multipara than nullipara
Race	Less common in Jewish and Muslim women
Sexual activity	Very rare in nuns
Age at first coitus	Common with first coitus at an early age[a]
Promiscuity	Commoner in those with a history of papillomavirus type 16 (wart virus)
Circumcision	Circumcision of male may be a protective factor for female
Social status	Commoner in lower socioeconomic groups
High-risk male	Higher in wives of unskilled labourers, and those who travel frequently
Oral contraceptive use	Commoner
Smoking	Commoner
Barrier contraception	Protective

[a] Correlation disappears when corrected for number of partners

non-specific erosion to a large tumour mass. Bleeding on touch is a characteristic sign. The differential diagnosis includes condyloma acuminata, amoeboma, tuberculosis and schistosomiasis. Cytology is often negative in invasive cancer due to surface slough.

General investigations include full blood counts, chest radiograph and intravenous pyelogram. Examination under anaesthesia with both vaginal and rectal palpation is performed to establish the staging (Table 26.1). It should be noted that the staging of cervical cancer is carried out before any operation, and often proves to be inaccurate. Some 15% of stage Ib lesions have positive pelvic nodes, and 6% have positive para-aortic nodes. The lesion is biopsied (with colposcopic assistance if necessary) and a cystoscopy performed to look for invasion (or the bullous oedema that suggests invasion) of the bladder.

Stages I and IIa may be treated by either surgery or radiotherapy or a combination of the two.

Surgery is appropriate for early stage disease. For FIGO stage Ia, if fertility is a requirement, the cone biopsy by which the diagnosis has been confirmed is sufficient. For stages Ib, IIa and some cases of IIb, radical surgery is appropriate. This removes the central tumour with the surrounding connective tissue (parametrium including cardinal and uterosacral ligaments) and draining lymph nodes. The operation can be a Wertheim hysterectomy or a Schauta radical vaginal hysterectomy.

Radiotherapy is used for cases in which surgery is contraindicated, in centres where surgical expertise is not available, or in advanced disease (large volume stage I, stages II–IV). The usual treatment is a course of external radiotherapy to the pelvis with additional intracavitary brachytherapy to the cervix and upper vagina. External radiotherapy (45–50 Gy) is delivered by a linear accelerator. Intracavitary brachytherapy (25–30 Gy) is usually delivered by a remote afterloading system such as Selectron using caesium-137 or indium-192. Radium is no longer used in the UK. The theoretical target point for brachytherapy is "point A", where the ureter crosses the uterine artery; this is the reference point for normal tissue tolerance. Point A is measured on the X-ray planning film as 2 cm lateral to and 2 cm below the intracavitary source.

A disadvantage of radiotherapy is that a definitive assessment of lymph node involvement is never possible. Complications of radiotherapy include damage to:

1. Skin (very rare with modern methods).
2. Vagina: stenosis occurs in 85% unless prevented by topical oestrogen treatment and vaginal dilators.
3. The bladder (an acute cystitis is common, necrosis, ulceration and formation of vesicovaginal fistula (3–8 months after treatment) are rare; and more likely to be due to tumour recurrence than the treatment.
4. The intestine (a transient enterocolitis is common; more severe necrosis with ulcer and fistula formation is rare; external radiation can cause extensive adhesions ("plastic" peritonitis).
5. Fractured neck of femur.
6. Leucopenia.
7. Exacerbation of pelvic inflammatory disease.

Recurrence after radiotherapy can be treated only by surgery or chemotherapy. Lateral recurrence with leg pain and oedema are untreatable except palliatively

Table 26.3. Published ranges of five year survival rates after treatment of carcinoma of the cervix. The worst figures are found (a) in young patients and (b) with larger primary tumours (more than 3 cm)

Stage	% survival
I	68–90
II	43–75
III	24–37
IV	0–14
All	41–59

by chemotherapy; even with treatment the remission rate is low (10% total; 30% partial). Central recurrence can be treated by surgery – usually anterior or posterior exenteration or both.

Surgical treatment of carcinoma of the cervix ranges from simple conisation for microinvasive disease to pelvic exenteration. Exenteration has a 50% 5-year survival if lymph nodes are not involved and 0% if they are. The operative mortality is 5% owing to pulmonary embolism and intestinal obstruction or perforation owing to adherence of the ileum to the denuded pelvic area. Rectovaginal fistula is also a common complication. The advantages of surgery over radiotherapy are: (a) the vagina, though shortened, is not narrowed or scarred; (b) the ovaries may be conserved; (c) a more accurate prognosis may be given; and (d) the side-effects of radiation (cystitis and proctitis) are avoided.

Some centres use combinations of surgery and radiotherapy. Five year survival rates for carcinoma of the cervix are shown in Table 26.3. Recurrence, if it occurs, is usually apparent within 18 months of treatment. The causes of death, in order of frequency, are uraemia, cachexia, haemorrhage, complications of treatment, and distant metastases.

Specific complications of radical surgery for carcinoma of the cervix include ureteric fistula and lymphocysts on the lateral pelvic wall. There is a higher primary mortality (0.5%) than with radiotherapy; fistulae (1%) and bladder dysfunction (10%) are more common.

Carcinoma of the Cervix in Pregnancy

This is a rare complication (1 in 5000–20 000 pregnancies) but is becoming more common as the disease affects younger patients. The prognosis is worse for the following reasons: (a) the disease follows a more aggressive course in young patients; (b) the lesion is more vascular; and (c) there may be delay between diagnosis and treatment. Treatment is an unsatisfactory balance between concern for the mother and concern for the child. Up to and including microinvasive lesions, the tendency is to delay definitive treatment until after delivery. With deeper invasion, treatment delay should be kept to a minimum as the chances for survival decrease by 2% per week between diagnosis and treatment. Treatment depends on gestation at the time of diagnosis. Before 20 weeks, early stage disease is treated by radical hysterectomy or radical radiotherapy (which results in abortion). After 28 weeks a caesarean Wertheim's hysterectomy is performed, or a classical caesarean section followed by radiotherapy.

Chapter 27
Benign and Malignant Lesions of the Endometrium

Benign Tumours of the Endometrium

Endometrial polyps consist of stroma and glands covered by a single layer of columnar epithelium. They are often part of a general endometrial hyperplasia. On their own they give rise to intermenstrual or postmenopausal bleeding. They are removed with sponge forceps or by curettage.

Carcinoma of the Endometrium

This has a maximum incidence at age 55 years. The tumour is usually a polypoid or nodular adenocarcinoma, commencing in the fundus. In approximately half of all cases the uterus is enlarged. The majority are adenocarcinoma, though 10% of cases have some malignant squamous elements (adenoacanthoma). The histology ranges from atypical hyperplasia to obvious neoplasia; many tumours show a spectrum of such changes. Spread is mostly direct in the endometrium. Invasion of the myometrium is less common. Lymphatic spread is uncommon and occurs to the iliac nodes and via the ovarian vessels to the para-aortic nodes. Still more rarely it may spread via the round ligament to the inguinal nodes. The ovaries and pouch of Douglas may be involved by transtubal embolism. Rare complications include pyometra. Grading of carcinoma of the endometrium is summarised in Table 27.1. Factors that may affect the incidence/aetiology of the condition are shown in Table 27.2.

The main clinical features are intermenstrual or postmenopausal bleeding (20% of cases present premenopausally). Immediate follow-up is a TVS for measurement of endometrial thickness; if this is less than 5 mm there is no malignancy. Definitive diagnosis is by endometrial biopsy and hysteroscopy. Cervical cytology yields positive findings in 25% of cases. Endometrial cytology is not reliable. Curettage has a 5–10% false-negative rate; therefore, hysteroscopy is always indicated. The treatment for stage I lesions is total hysterectomy with a cuff of vagina and bilateral salpingo-oophorectomy or laparoscopic assisted vaginal hysterectomy. Radiotherapy alone is much less satisfactory than surgery, though

Table 27.1. Staging of carcinoma of the uterine corpus (FIGO 1988)

Stage	
IA G123	Tumour limited to endometrium
IB G123	Invasion to < 1/2 myometrium
IC G123	Invasion > 1/2 myometrium
IIA G123	Endocervical glandular involvement only
IIB G123	Cervical stromal invasion
IIIA G123	Tumour invades serosa and/or adnexa and/or positive peritoneal cytology
IIIB G123	Metastases to pelvic and/or para-aortic lymph nodes
IVA G123	Tumour invasion of bladder and/or bowel mucosa
IVB	Distant metastases including intra-abdominal and/or inguinal lymph nodes

Histopathology: degree of differentiation
Cases of carcinoma of the corpus should be grouped with regard to the degree of differentiation of the adenocarcinoma as follows:
G1 5% or less of non-squamous growth pattern
G2 6–50% of a non-squamous growth pattern
G3 More than 50% of a non-squamous growth pattern

Notes on pathological grading
1. Notable nuclear atypia, inappropriate for the architectural grade, raises the grade of a grade I or grade II tumour by one
2. In serous adenocarcinomas, clear cell adenocarcinomas, and squamous cell carcinomas, nuclear grading takes precedence
3. Adenocarcinomas with squamous differentiation are graded according to the nuclear grade of the glandular component

Table 27.2. Factors that may affect the incidence/aetiology of carcinoma of the endometrium. In addition, there is a group of factors whose importance, though often discussed, is in considerable doubt (e.g. age at menopause, radium menopause, diabetes and hypertension)

Factor	Effects
Age	Commoner in older women, maximum incidence at 60 years
Parity	More frequent (2–3 times) in nullipara (but most patients are parous)
Obesity	Commoner in overweight subjects associated with peripheral production of oestrone
Hormone replacement therapy	With unopposed oestrogens[a]
Granulosa cell tumour	Unopposed oestrogens
Endometrial polyps	
Persistent juvenile menorrhagia	Unopposed oestrogens
Polycystic ovarian syndrome	

[a] Combined cyclical oestrogen and progesterone treatment protects against cancer of the endometrium.

postoperative external irradiation is administered if invasion extends more than one third through the myometrial wall, or the histology is anaplastic or mixed müllerian. The place of pelvic lymph-adenectomy is controversial. The major site of recurrence is in the vaginal vault and the incidence of this can be reduced from 10% to 1% by postoperative vault radiotherapy. The treatment for stages II and III lesions is total hysterectomy, bilateral salpingo-oophorectomy and radiotherapy. The 5 year survival rate is: stage I, 75–80%; stage II, 55–60%; stage

III, 35–40%; stage IV, less than 10%. High dose progestogen therapy is advocated as palliative treatment for recurrent or otherwise untreatable tumours, and is said to produce temporary amelioration of bone pain and to produce regression of pulmonary metastases.

Chapter 28
Benign and Malignant Tumours of the Myometrium

Benign Tumours of the Myometrium: Fibroids

Leiomyoma (fibromyoma, fibroids) is the commonest of all tumours in the human body (20% incidence at post mortem). Factors affecting the incidence include parity (commoner in nullipara) and race (commoner in black women, but with no association with parity).

The tumour may lie entirely in the uterine wall (intramural), or project through the peritoneal surface (subserous), or become pedunculated and even entirely separate from the uterus (parasitic). The tumour commonly arises in the body of the uterus, less commonly in the cervix, and very rarely in the round ligament. It may be single or multiple. Microscopically they consist of whorls of smooth muscle cells. Large tumours have a greater proportion of fibrous tissue and because of their relative avascularity may show various types of degeneration – necrosis, hyaline change, calcification and cystic areas. Infection may occur at the tip of an infarcted fibroid polyp, especially after termination of pregnancy. Malignant change is rare (0.5%).

There may be no symptoms, especially after the menopause. Specific clinical features may include abdominal swelling, pressure effects: oedema, varicose veins, haemorrhoids, disturbance of micturition and menorrhagia. Very large fibroids may cause metabolic effects, hypoglycaemia and polycythaemia. The severity of menorrhagia is often unrelated to the size and position of the tumours. However, submucous fibroids are particularly likely to cause menorrhagia and fibroid polyps to cause intermenstrual bleeding. Pain may occur due to congestion or backache due to retroversion; more severe pain results from degeneration, torsion of a pedunculated fibroid, or expulsion of a fibroid polyp. The differential diagnosis of fibroids includes pelvic masses in general and adenomyoma, though they can usually be distinguished from extrauterine pathology. Fibroids are associated with infertility.

Symptomless fibroids usually require no treatment. If they are causing symptoms, or the diagnosis is in doubt, or they are larger than a 16-week pregnancy then surgery is indicated. Hysterectomy is the procedure of choice. If this is contraindicated by the age or wishes of the patient, or the desire for further pregnancies, myomectomy can be performed. However, this operation has a higher morbidity than hysterectomy (postoperative bleeding and intestinal obstruction due to adhesion of bowel to the uterine scar). Many patients who have a myomectomy will eventually require a hysterectomy for recurrence. Prior to surgery, it is possible to reduce the size of fibroids and to relieve symptoms by the use of GnRH agonists.

Fibroids During Pregnancy

Fibroids may enlarge and soften during pregnancy, and involute during the puerperium. Infarction may occur and lead to "red degeneration". This gives rise to abdominal pain, tenderness and signs of peritoneal irritation; temperature and pulse may be slightly raised. Treatment is conservative. Torsion of a pedunculated fibroid may occur, especially in the early puerperium; treatment is by laparotomy and myomectomy.

In early pregnancy abortion is commoner, especially with submucous fibroids. Fibroids rarely cause mechanical problems during labour, though very rarely a cervical tumour can lead to non-engagement of the fetal head or even obstructed labour. Management of the latter situation is by caesarean section but not myomectomy. Fibroids can interfere with retraction of the uterus and thus be responsible for postpartum haemorrhage.

Metastasising Fibroids

These conditions are rare and represent behaviour intermediate between that of local benign fibroids and unequivocal sarcomas.

Histologically the tumour is indistinguishable from a benign fibroid but metastasis occurs in one of two forms. The first is haematogenous spread with one or more secondary growths, usually in the lung. These seldom recur after local removal and the prognosis is excellent. Cords of tumour tissue may also extend along blood vessels. The second form is transperitoneal spread, which superficially resembles ovarian cancer.

Malignant Non-epithelial Tumours of the Uterus

These are rare but highly malignant. About 50% are leiomyosarcomas arising from normal myometrium or fibroids. Less common types arising from the endometrial stroma include stromal sarcoma, fibromyxosarcoma, carcinosarcoma and mesodermal mixed tumours (embryonal sarcoma). The latter are found in the cervix and vagina as well as the corpus and have a "grape-like" appearance (sarcoma botryoides). These have a loose myxomatous stroma within which may be found bone, cartilage and striped muscle. Other rare tumours include angiosarcoma, haemangiopericytoma and lymphosarcoma.

Microscopically they are similar to leiomyomas but with pleomorphic features characteristic of malignancy. Lymphatic spread is uncommon and metastasis is usually haematogenous. The commonest age group for leiomyosarcomas is 40–60 years; two thirds are postmenopausal. The symptoms are bleeding, pain and a

Table 28.1. A classification of endometrial stromal tumours

Classification	Features
Stromal endometriosis	Consists of stromal endometrial cells with little pleomorphism and few mitoses. Slow spread; low mortality and recurrences
Endometrial stromal sarcoma	Similar histology to above but with greater pleomorphism and more mitoses. Much worse prognosis
Mixed müllerian tumour	This consists of a variety of sarcomatous tissues such as stromal sarcoma, rhabdomyosarcoma. Carcinomatous elements may also be present (carcinosarcoma). The prognosis is very poor

rapidly expanding pelvic mass or features of lung or bone metastasis. Treatment is by hysterectomy and salpingo-oophorectomy. Some use radiotherapy or chemotherapy, especially for palliation.

The commonest age group for endometrial "stromal" sarcomas is 50–70 years. A classification of endometrial stromal tumours is shown in Table 28.1. Sarcoma botryoides usually occurs in children under 3 years. These tumours are fairly sensitive to radiotherapy and chemotherapy (triple therapy with vincristine, actinomycin D and cyclophosphamide). Definitive surgical treatment consists of extended hysterectomy and total vaginectomy, though vaginal sarcoma botryoides in children is never treated solely with chemotherapy and has a good prognosis.

Chapter 29

Tumours of the Ovary

Benign Tumours of the Ovary

These include the following:

1. *Mucinous cystadenoma*: large, multilocular, unilateral cysts with mucinous contents, lined by a single layer of columnar cells. Papillary processes are rare and suggest malignancy. Malignant change is present in 5–10% of cases of mucinous cystadenocarcinoma. A rare complication is pseudomyxoma peritonei in which the cells implant throughout the peritoneum and continue to secrete mucin.
2. *Serous cystadenoma*: the incidence is one quarter that of mucinous cystadenomas. They are smaller than mucinous lesions, bilateral in 50% of cases, often unilocular, and lined by cubical or columnar epithelium (sometimes ciliated). Papillary processes are common but are likely to be malignant only if they penetrate the capsule. The stroma may contain small calcium deposits ("psammoma bodies").
3. *Dermoid cysts*: these are the commonest neoplastic cysts in young women but may be found at any time of life. They are usually less than 12 cm in diameter; 10% are bilateral. Cysts lying in the pouch of Douglas or anterior to the uterus are usually dermoids. The contents are thick and yellow. The main constituent is skin and associated structures (hair, sebaceous and sweat glands, and teeth) but a variety of other tissues may be found (alimentary and respiratory epithelium); functional thyroid tissue (struma ovarii); neural tissue (cartilage and bone); very rarely, there may be partial embryo formation (fetus in fetu). Solid teratomata may be benign but are more commonly malignant. Primary choriocarcinoma, melanoma and carcinoid can occur. Unlike gastrointestinal argentaffinomas, ovarian carcinoid may cause systemic effects prior to metastasis because the serotonin is not destroyed in the enterohepatic circulation.
4. *Fibroma*: these are rare, lobulated tumours, bilateral in 10% of cases. They often produce ascites and occasionally a right-sided pleural effusion (Meig's syndrome). They may be difficult to distinguish macro- and microscopically from theca cell tumours.

With the exception of dermoids, the incidence of benign ovarian tumours rises with age. Benign tumours are often symptomless (especially dermoids) but may be present with abdominal enlargement. Dermoid cysts are often diagnosed in pregnancy as an incidental finding at ultrasound examination. Pressure effects include lower-limb oedema, varicosities, haemorrhoids, and frequency or urgency of micturition, but not constipation; the very largest tumours may cause dyspnoea or even cachexia by interfering with digestion. On examination, a large ovarian cyst is dull to percussion anteriorly but resonant in the flanks. Diagnosis can be confirmed by ultrasound and laparoscopy. A variety of accidents to ovarian tumours may occur and give rise to acute or subacute abdominal symptoms: these include torsion (usually a benign cyst of moderate size with a long pedicle), rupture (contents of a haemorrhagic or dermoid cyst can lead to severe irritation), haemorrhage, and infection (can occur with chocolate or malignant cysts adherent to the bowel).

Excision by laparotomy or laparoscopy is usually recommended for any tumour more than 5 cm in diameter. However, if the cyst is multilocular or semisolid on ultrasound scan it should be removed regardless of size. Thin-walled cysts are usually physiological and can be monitored by serial ultrasound. This reduces the need for ovarian cystectomy, which can lead to fimbrial damage. In younger women treatment is enucleation or excision of the cyst with conservation of ovarian tissue if possible. In older women oophorectomy is performed. If both ovaries are removed (e.g. bilateral large cysts) hysterectomy should also be carried out.

Special Tumours of the Ovary

Feminising Tumours

Granulosa and theca cell tumours are the commonest of the special group. Granulosa cell tumours, the more frequent of the two, can arise at any age and are solid, lobulated, yellow and usually unilateral. The cells are in sheets grouped around spaces suggestive of primordial follicles. In theca cell tumours spindle-shaped cells are arranged in intertwining bundles. Some tumours may show both granulosa and theca elements, and all types can show luteinisation. All of the ovarian steroid-secreting tumours are thought to arise from gonadal stromal cells. Typically, but not invariably, the granulosa and theca cell tumours secrete oestrogens and according to the patient's age group may cause precocious puberty, dysfunctional bleeding due to endometrial hyperplasia or post-menopausal bleeding. Around half of all granulosa cell tumours are malignant and may present as a painful mass or ascites. Theca cell tumours are benign. Recurrences may occur at very long intervals after the original treatment and may require repeated debulking operations. Histology provides little or no clue as to the malignant potential. There is an association between feminising tumours and endometrial carcinoma.

Masculinising Tumours

Masculinising tumours or arrhenoblastoma (androblastoma), usually occur in younger women as a small, unilateral cystic mass. The tumour consists of Leydig-like cells. They may be highly differentiated with structures comparable to testicular tubules (tubular adenoma of Pick) ranging to an undifferentiated sarcomatous appearance. "Reinke" crystals are an important microscopical feature. Typically they secrete androgens leading to oligo-and amenorrhoea, hirsutes and virilisation, breast atrophy, clitoral hypertrophy and deepening of the voice. Around one-quarter are malignant. Other varieties include a mixed granulosa-cell and arrenhoblastoma (gynandroblastoma), and a group of androgen-secreting tumours with characteristic lipid-containing cells.

Malignant Germ Cell Tumours (Table 29.1)

1. *Disgerminoma*: this is the commonest germ cell tumour, consisting of germ cells identical with the testicular seminoma, with sheets of large polyhedral cells. Some 10% show mixed teratomatous elements. The tumour is commonest at age 20–30 years, and consists of a lobulated solid mass, bilateral in 20–30%. It is also the commonest ovarian malignancy in childhood and adolescence. It usually presents as a mass, with or without pain. Most, but not all, germ cell tumours produce hCG or AFP or both. Those with elements of endodermal sinus, and embryonal carcinoma, may secrete AFP. If unilateral it can be managed by simple oophorectomy. Advanced or recurrent tumour is highly sensitive to chemotherapy, the commonest combination being vincristine, actinomycin D and cyclophosphamide (VAC). Overall long-term survival of this previously lethal malignancy is now 85%.
2. *Malignant teratoma*: these are usually solid teratomas, one third of which are malignant. Early blood spread is characteristic.
3. *Choriocarcinoma*.
4. *Endodermal sinus tumour* (yolk sac tumour): this is the most rapidly lethal tumour known but now has excellent prognosis with early treatment. Most of these tumours secrete AFP.

Table 29.1. Classification of malignant germ cell tumours of the ovary

A. Disgerminoma
B. Endodermal sinus tumour
C. Teratomas
   1. Immature (malignant teratoma)
   2. Mature cystic teratoma with malignant transformation
   3. Monodermal or highly specialised:
      a. Struma ovarii
      b. Carcinoid
      c. Strumal carcinoid
      d. Others
D. Embryonal carcinoma
E. Choriocarcinoma
F. Combination germ cell tumour

Brenner Tumours

The tumour comprises nests of transitional epithelial cells in a fibrous stroma, sometimes with cysts suggestive of Graafian follicles. It may be found in the wall of a mucinous cyst. As some cases are malignant it should be treated by hysterectomy and bilateral salpingo-oophorectomy, though in younger women with a strictly localised and unilateral lesion a more conservative approach is possible.

Malignant Epithelial Tumours of the Ovary

These are usually found in older women. Associated factors are listed in Table 29.2 and staging and classification in Tables 29.3 and 29.4. The commonest form is serous cystadenocarcinoma (35–50%); mucinous cancers account for 6–10%. The tumour may be solid or cystic or both, and can arise in a benign cyst, though malignancy ab initio is more typical. Over 50% are bilateral. Direct invasion of all the surrounding pelvic organs is the most prominent form of spread and peritoneal involvement leads to ascites. Lymphatic spread occurs to the uterus, tubes, opposite ovary, para-aortic lymph nodes and, in terminal cases, the supraclavicular nodes. Haematogenous spread may involve the lung but other distant metastases are rare.

Secondary cancer of the ovary is usually bilateral and may arise by blood, lymphatic or direct spread from carcinoma of the endometrium, the stomach and large bowel, and the breast. An unusual type of secondary from stomach (or bowel or breasts) is the Krukenberg tumour; the primary may be very small whereas the ovarian lesion is a large, solid tumour with characteristic mucin-containing "signet ring" cells.

The clinical features of malignant ovarian tumours are a pelvic mass and pain, often not observed until spread has occurred. Cachexia is a late event, and abnormal bleeding is unusual. The tumour is usually immobile; deposits in the pouch of Douglas and ascites are common. Circulating levels of the ovarian tumour antigen CA 125 may be elevated in cystadenocarcinoma. A full diagnosis can only be made after surgical exploration and appropriate biopsies. The primary treatment for disease confined to one or both ovaries is hysterectomy and bilateral salpingo-oophorectomy and omentectomy. For tumours more advanced than stage I the treatment is hysterectomy, salpingo-oophorectomy, omentectomy and debulking, followed by chemotherapy. Typical survival rates are shown in Table 29.5. Operations may be followed by chemotherapy; this may be used even when

Table 29.2. Factors affecting the incidence/aetiology of ovarian cancer

Factor	Effects
Parity	Commoner in low parity, especially with women who have had no children
Oral contraceptives	Have a protective effect
Infection with mumps virus	Protective effect
Talc (dusting powders) containing asbestos	Higher incidence
Familial	–
Social class	Mortality higher in social class I than class V

Table 29.3. Staging of ovarian carcinoma (FIGO 1988)

Stage	
I	Growth limited to the ovaries
Ia	Growth limited to *one* ovary; no ascites; no tumour on the external surface; capsule intact
Ib	Growth limited to *both* ovaries; no ascites; no tumour on the external surface; capsule intact
Ic	Tumour either stage Ia or Ib, but *with tumour* on external surface of one or both ovaries; *with ascites* present containing malignant cells or with positive peritoneal washings
II	Growth involving one or both ovaries with pelvic extension
IIa	Extension and/or metastases to the *uterus and/or tubes*
IIb	Extension to other *pelvic tissues*
IIc	Tumour either stage IIa or IIb but with tumour on the surface of one or both ovaries; or with capsule(s) ruptured; or with ascites present containing malignant cells or with positive peritoneal washings
III	Tumour involving one or both ovaries with peritoneal implants outside the pelvis and/or positive retroperitoneal or inguinal nodes; superficial liver metastases equals stage III; tumour is limited to the true pelvis but with histologically proven extension to the small bowel or omentum
IIIa	Tumour grossly limited to the true pelvis with *negative nodes* but with histologically confirmed *microscopic seeding* of abdominal peritoneal surfaces
IIIb	Tumour involving one or both ovaries with histologically confirmed implants of abdominal peritoneal surfaces *none exceeding 2 cm in diameter; nodes are negative*
IIIc	Abdominal implants *greater than 2 cm in diameter and/or positive retroperitoneal or inguinal nodes*
IV	Growth involving one or both ovaries with distant metastases. If pleural effusion present, there must be positive cytology to allot a case to stage IV; parenchyma liver metastases equals stage IV

To assess the impact on prognosis of the different criteria for allotting cases to stage Ic or IIc it would be of value to know whether the source of malignant cells was (a) peritoneal washings or (b) ascites, and whether rupture of the capsule was spontaneous or caused by the surgeon

Table 29.4. Histological classification of tumours of the ovary

I Serious cystomas
   A. Serous benign cystadenoma
   B. Serous cystadenoma with proliferative activity of the epithelial cells and nuclear abnormalities but with no infiltrative growth (borderline)
   C. Serous cystadenocarcinoma

II Mucinous cystomas
   A. Mucinous benign cystadenoma
   B. Mucinous cystadenoma (borderline)
   C. Mucinous cystadenocarcinoma

III Endometroid tumours (similar to adenocarcinoma in the endometrium)
   A. Endometrioid cysts
   B. Endometrioid cysts (borderline)
   C. Endometrioid adenocarcinoma

IV Concomitant carcinomata. Unclassified carcinomata (tumours which cannot be allotted to one of the above groups)

Table 29.5. Typical five year survival rates for ovarian cancer[a]

Stage	%	
Ia	90	
Ib	64	66
Ic	50	
IIa	52	45
IIb and IIc	42	
III	20–30	10
IV	< 10	

[a] Despite the depressing five year survival figures for stages III and IV, many patients (~50%) will have a useful remission of 2–3 years with chemotherapy. A further 30% have a partial remission. The response rate for further chemotherapy after a recurrence is, however, much worse.

there is no demonstrable disease after operation ("adjuvant" chemotherapy). Chemotherapy is usually a combination of cytotoxic drugs including a platinum-based compound and a taxane. Overall survival is poor (30%), largely because of late diagnosis. The prognosis following surgery is related directly to the extent of residual disease, irrespective of the total volume of tumour before resection. The greater the interval between the primary treatment and recurrence, the better the prognosis.

Chapter 30

Infertility

This is defined as inability to conceive for 12 months, during which approximately 80–90% of couples desirous of pregnancy will have done so. The prevalence is 14%, at least half of couples complaining of secondary infertility. The infertility will remain unresolved in 3–4% of couples. The mean time to pregnancy rises from 6 months in women at all ages attempting their first pregnancy to 15 months in a 35 year old undergoing artificial insemination with donor sperm (DI) or more than 7 years if, in addition, this patient has been treated for endometriosis. The monthly probability of pregnancy (fecundability) expressed as the reciprocal of the mean time required for conception in months falls from 0.4 in the 25 year old attempting her first pregnancy to 0.008 after 2 years of unexplained infertility.

Aetiology

The major categories are ovulatory disorders (25%), sperm defects (25%), unexplained infertility (25%), tubal disease (20%), and endometriosis (5%). About 30% of couples have more than one cause identified.

Management

A history and physical examination of both partners is essential. In the male, information about previous testicular or inguinal surgery, infections, trauma and erectile/ejaculatory dysfunction should be sought. Normal testicular volume (> 20 ml) should be confirmed. Regular (24–36 days) menstrual cycles largely exclude ovulatory disorders. Pelvic surgery, infection or intrauterine contraceptive device usage may indicate tubal pathology. Coital frequency and normal pelvic anatomy should be confirmed.

Investigations

Semen analysis should be repeated two times, each after three days abstinence. Volume of 2–5 ml, > 20 million/ml density, motility > 40% and < 40% abnormal forms are generally considered to be normal. A biphasic basal body temperature chart indicates that ovulation has taken place. A luteal phase plasma progesterone level > 30 nmol/ml confirms that ovulation has occurred. Pulsatile LH secretion reaches a peak 36 h before ovulation. This information may be used to time sexual intercourse. A normal ovum can be fertilised only 12–24 h after release, whereas sperm remain fertile for up to 72 h. The uterine cavity, tubal patency and pelvic normality are assessed by hysterosalpingography, hysterosalpingocontrast sonography (HyCoSy), hysteroscopy and/or laparoscopy and dye injection.

The significance of sperm agglutination and sperm immobilising antibody tests is controversial. Inadequate corpus luteum function (as reflected by persistently low progesterone levels after ovulation) is thought by some to be a cause of infertility. Even if it does exist, treatment with progesterone, hCG or pituitary gonadotrophins is of doubtful value.

Treatment

If all tests are normal after 2 years of infertility (so-called unexplained infertility), assisted conception techniques should be used. Medical and surgical treatment of male factor infertility is usually disappointing. Viable options include intrauterine insemination of washed prepared sperm and assisted conception techniques with or without donor sperm, including intracytoplasmic sperm injection (ICSI) into oocytes in vitro. Sperm may be obtained by masturbation or needle aspiration from the epidydimis or the testis by needle aspiration or open surgery. In men with severe oligoasthenospermia or obstructive azoospermia, screening for cystic fibrosis and chromosomal abnormalities, in particular microdeletions of the Y chromosome, is recommended. Given the success of ICSI in the treatment of male factor infertility, the two most important prognostic factors are the age of the woman and the duration of infertility.

The treatment of failure of ovulation has been described and typically includes the use of oral clomiphene citrate (50–150 mg daily for 5 days in the follicular phase) or if unsuccessful, pituitary gonadotrophins (hMG or rec FSH) (injected daily or less frequently according to the cause of anovulation). Monitoring of these regimens includes serial transvaginal ultrasound to observe follicular development and serum oestradiol measurements to reduce the risks of multiple pregnancy and ovarian hyperstimulation syndrome (0.2–10%). Tubal microsurgery is advocated by some as first-line treatment for tubal infertility. Time-specific intrauterine pregnancy rates up to 60% after 3 years for salpingolysis and fimbrioplasty, and 25–30% after salpingostomy for distal tube occlusion have been claimed. Prognosis is influenced by the type and extent of the disease as well as the quality of surgery. Others prefer in vitro fertilisation/embryo transfer (IVF-ET) particularly in the most severe cases of tubal disease. Pregnancy rates per cycle for women under 40 years are 25–30%, with a cumulative "take-home baby" rate up to 60% after four cycles of IVF-ET. If intrauterine insemination (IUI) is preferred in "unexplained infertility", it must be performed with superovulation; pregnancy rate is 25% per cycle and the cumulative rate is 69% after six cycles. Thereafter, IVF-ET is advised. The use of gamete intrafallopian tube transfer (GIFT) has declined in the UK since the success rates with IVF-ET are equivalent and IVF-ET provides useful diagnostic information to determine subsequent treatment. In the presence of seminal antisperm antibodies, IVF-ET is the preferred strategy; the apparently lower fertilisation rates do not affect pregnancy rates of 25–30% per treatment cycle. The use of donor oocytes are indicated in premature ovarian failure or in women with a poor response to ovarian stimulation.

Chapter 31

Contraception

Approximately 50% of women aged 18–44 years in the UK use reversible contraception and 50% of these rely on the oral contraceptive pill, 25% use condoms and 12% the IUCD. About 8% rely on withdrawal (coitus interruptus), the remainder using the diaphragm with or without spermicides or the "safe period" (rhythm method). The effectiveness of all these methods is measured using the Pearl Index – the number of unwanted pregnancies that occur during 100 woman-years of exposure (normally fertile women having regular coitus). The natural pregnancy rate (i.e. the Pearl Index without any birth control) is approximately 80. Hormone replacement therapy does not provide adequate contraceptive protection.

Natural Family Planning

Coitus Interuptus

Coitus interuptus relies on withdrawal of the penis before ejaculation, so the effectiveness depends very much on the technique and enthusiasm of the male partner. It is widely and successfully used in situations where there is no access to other contraceptives (failure rate 8–17 per 100 woman-years).

Fertility Awareness

The rhythm method (also called fertility awareness and natural family planning) involves predicting the date of ovulation and avoiding intercourse 7 days before and 3 days after (days 7–16 in a 28 day cycle). The calendar method relies on monitoring the cycle for 12 months and using the longest and shortest cycles to predict the likely fertile period. Other factors that can be used to help pinpoint the time of ovulation include the postovulatory rise in basal temperature, mittelschmertz (ovulation pain), and changes in cervical mucus. Biochemical measurements such as urinary lutenising hormone and oestrone 3-glucuronide can be used and are combined in the 'Persona' system, which uses a microcomputer to calculate the safe days. The failure rate is 2–20 per 100 woman-years.

Barrier Methods

Condoms

Condoms are made of thin vulcanised latex rubber and are usually lubricated, sometimes with spermicide. If used properly they can protect against sexually transmitted diseases. Problems include reduction in male sensation, allergy to spermicides and breakage, which may go unnoticed. Quoted failure rate 2–15 per 100 woman-years.

Diaphragms and Caps

The diaphragm consists of a latex rubber hemisphere in various sizes, surrounded by a spring to keep it in shape. The flat spring is the most commonly used, the coil spring is softer but more difficult to insert and the arcing spring forms an arc when compressed, which can make it easier to insert. When in place it rests between the retro pubic ledge and the posterior fornix of the vagina. It is inserted by the woman prior to intercourse and removed 6 h later. Its use requires some skill in identifying the cervix and ensuring that it is covered, so the failure rate declines with duration of use. The quoted failure rate is 4–18 per 100 woman-years. The diaphragm functions as a carrier for the spermicide and a barrier between the acidic vagina, which is hostile to sperm, and the alkaline cervical mucus, which is receptive to sperm. It gives some protection against cervical STDs (chlamydia and gonorrhoea), but not others. It can cause cystitis or vaginal soreness due to allergy or reaction to the spermicide. Cervical and vault caps have similar features, but they are kept in position by suction rather than spring tension.

Spermicides

Spermicides are substances that destroy sperm (e.g. nonoxynol 9), and are best used with barrier methods. The contraceptive sponge, impregnated with spermicide, has been withdrawn from use because of high failure rates. Quoted failure rates for spermicide alone 4–25 per 100 woman-years.

The Female Condom

The female condom (femidom) is a soft, loose fitting polyurethane sheath with a loose inner ring to hold it across the vaginal fornices, and an outer ring outside the vaginal introitus. It is available over the counter but is not widely used. It protects against STDs, and the failure rate is 5–15%.

The Intrauterine Contraceptive Device

The IUCD may be inert or contain an active ingredient (e.g. copper (Multiload) or a progestagen (e.g. Mirena). The IUCD is most suitable for multiparae who have completed their family. In addition to stimulating a foreign body reaction in the endometrium, the IUCD impairs sperm migration, fertilisation and ovum transport. Copper is spermicidal. The failure rate for IUCDs is 0.3–2 per 100 woman-years.

Absolute contraindications include malignant disease of the genital tract, vaginal bleeding of unknown cause, suspected pregnancy, active PID, previous ectopic pregnancy and Wilson's disease. Relative contraindications include nulliparity, recent history of PID, cervical stenosis, valvular heart disease and septal defects. The insertion of the IUCD carries a small but definite risk of infective endocarditis; antibiotic prophylaxis is required in susceptible patients. Higher expulsion rates occur during or immediately after menstruation as well as in the first eight postpartum weeks. The IUCD should be continued until 1 year after the menopause.

Complications include the following:

1. Menorrhagia, intermenstrual bleeding and dysmenorrhoea; these improve with use.
2. Missing IUCD, which may be (a) drawn up into the uterine cavity, (b) perforating the uterus (0.1–1% of devices), or (c) expelled through the vagina. Pregnancy should be excluded and the device located using ultrasound or abdominal radiograph. Extrauterine devices should be removed by laparoscopy or laparotomy. There is an increased risk of abortion but not fetal abnormality.
3. A risk of PID, which is highest in the first few months after insertion (relative risk for PID) in IUCD users is 1.5–2.6). The incidence of PID in IUCD users under the age of 20 years is more than 10 times that of 30–49 year olds. Treatment includes removing the device and administering antibiotics. Vaginal actinomycosis is commoner in IUCD users. Women should be prescreened for chlamydia.
4. A marginal increase in tubal infertility in nulliparous but not parous users of the copper IUCD.
5. Ectopic pregnancy occurs in 3–9% of all pregnancies that occur with an IUCD in place, perhaps because IUCDs provide more protection against intrauterine than against tubal pregnancies. The lowest rates of ectopic pregnancy are associated with copper devices and the highest with progesterone devices. The commonest symptoms of early ectopic pregnancy in an IUCD user are either a missed period or a very light period with or without pelvic pain. Patients should be instructed to feel for the device monthly and attend 6-monthly check-ups. Mirena needs to be replaced every 3 years and copper devices every 3–8 years.
6. There is a 25% risk of miscarriage, rising to 50% if the IUD is removed.
7. Perforation of the uterine wall (1 in 5000 cases).

Hormonal Contraception

The Combined Oral Contraceptive Pill

This is used by 20–30% of all sexually active women. The combined oral contraceptive pill (COCP) contains oestrogen (oestradiol or mestranol, 20–50 μg daily) and progesterone (19 norsteroids such as norethisterone, norgestrel, ethynodiol, lynestrenol; and the newer generation compounds such as desogestrel, gestodene and norgestimate). Two types of preparations are available: (a) fixed-dose pills (monophasic) in which oestrogen and progesterone are taken in constant doses for 21 days followed by a 7-day pill-free period; and (b) multiphasic preparations that contain variable doses of progesterone and oestrogen and allow the use of the lowest dose of both hormones to provide good cycle control and least side-effects. Both oestrogen and progesterone combine to suppress ovulation (by suppressing both LH and FSH secretion) and interfere with synchronised endometrial development; progesterone renders the cervical mucus scanty and viscous, thus impairing sperm transport.

Dose-related metabolic effects of the COCP include alterations in liver function, reduced glucose tolerance, increased levels of triglycerides and low density lipoprotein (LDL) cholesterol but decreased high density lipoprotein (HDL) cholesterol levels (i.e. an unfavourable lipid profile). The new generation progestagens (e.g. desonorgestrel) cause fewer metabolic effects and lead to an elevation in HDL cholesterol and a reduction in LDL cholesterol. The COCP induces a hypercoagulable state, by increasing plasma fibrinogen and the activity of factors VII and X, while reducing antithrombin III levels. Many drugs interfere with the contraceptive effect of the COCP including non-steroidal anti-inflammatory agents, anticonvulsants, antihistamines, antibiotics and major tranquillisers. Women on liver enzyme-inducing drugs (e.g. phenytoin, rifampicin) may need to increase to a higher oestrogen pill if breakthrough bleeding occurs. Absolute contraindication to the COCP include pregnancy, pulmonary embolism, deep venous thrombosis, sickle cell disease, porphyria, liver disease and jaundice. Diabetics on the COCP have an increased risk of ischaemic heart disease and cerebrovascular accidents, but less than a quarter require adjustments insulin dosage. The pill should be discontinued if any of the following occur: severe headaches, visual disturbance, migraine for the first time and transient neurological changes.

Apart from reducing the incidence of intrauterine pregnancy (failure rate 0.2–0.5 per 100 woman-years), the COCP reduces risk of ectopic pregnancy by 90%. Similarly, there is a 40–50% reduced risk of PID (but not herpes or chlamydia). Fibroids, functional ovarian cysts and benign breast disease (fibroadenoma and fibrocystic disease) are less common after COCP use. The rates of carriage of *Candida albicans* are similar in pill users as in those using an IUCD. The COCP does not appear to protect against the development of rheumatoid arthritis, but it may prevent progression to more severe disease. There is a three- to sixfold increase in the relative risk of spontaneous venous thromboembolism in young women taking the pill, which is unrelated to duration of use. The newer progestational agents (e.g. desonorgestrel) may be associated with a slightly increased incidence of thromboembolism. The COCP should be stopped 4–6 weeks before major orthopaedic, abdominal or cancer surgery. On balance, the risk to the young woman on the pill of becoming pregnant from stopping the pill or of developing side-effects from heparin prophylaxis may be greater than the risk of developing postoperative deep venous thrombosis. The overall risk of myocardial infarction is increased about three to five times in COCP users and that of haemorhagic stroke is approximately doubled. The risk is highest in smokers over the age of 35 years. The overall mortality rate from circulatory causes in smokers below the age of 35 years taking the COCP is 1 in 10 000, increasing to 1 in 2000 in smokers aged between 35 and 44 years and 1 in 550 above the age of 45. Smokers over the age of 35 should not use the COCP. The COCP doubles the risk of mild hypertension (2.5% incidence over 5 years of use).

The risk of hepatocellular adenoma and carcinoma is increased in older women using high dose COCP for prolonged periods; benign liver tumours occur in 1 in 10 000 users. By contrast, the COCP has no effect on breast cancer risk in women aged between 25 and 39 years, though conclusions on the effect of COCP use both early and late in reproductive life are still controversial. In any case the lowest effective dose of oestrogen and progesterone should be used. The frequency of ovarian cancer and endometrial cancer is reduced. The reduction in risk persists for a least 10 years after stopping the COCP. The frequency of gall bladder disease is increased in women on the COCP.

Postcoital Contraception

The Yuzpe regimen consists of two high oestrogen pills (50 μg of ethinyloestradiol and 250 μg of levonorgestrel each) taken within 72 h or unprotected mid-cycle exposure and repeated after 12 h. Vomiting often necessitates repeating the treatment. The failure rate is 1-2% per cycle. It does not affect clotting factors and there are no published reports of death or serious illness after its use. The only contraindications are thromboembolism and pregnancy. An alternative is to insert the IUCD within 5 days of unprotected intercourse. The use of a progesterone antagonist (mifepristone) will increase in the future.

Progesterone-Only Pills

Progesterone-only pills (POP) contain either ethynodiol or norethisterone. These alter endometrial maturation and affect cervical mucus. They need to be taken at the same time every day (+3 h) for 3 weeks of a 4 week cycle. The failure rate is 1-4 pregnancies per 100 woman-years. Absolute contraindications include suspected pregnancy, undiagnosed vaginal bleeding, previous ectopic pregnancy, undiagnosed vaginal bleeding, previous ectopic pregnancy, following a hydatidiform mole (at least until hCG levels are normal), carcinoma of the breast and severe arterial disease. The incidence of amenorrhea, irregular breakthrough bleeding and functional ovarian cysts is increased.

Injectable Steroids

Injectable compounds consist of either intramuscular depot preparations such as medroxyprogesterone acetate, a 17α-hydroxyprogesterone derivative (Depo-Provera, 150 mg every 3 months) or norethisterone (Noristerat, 200 mg every

2 months). Side-effects of injectable steroids include amenorrhoea (up to 45% in the first year), irregular vaginal bleeding and weight gain. Menstrual cycles and fertility return to normal within 6 months of the last injection. These preparations do not increase the risk of carcinoma of the breast, endometrium or cervix.

The Norplant system for subdermal delivery of the synthetic progestin levonorgestrel has been introduced as a long-term contraceptive. The drug is sealed into six 2.44 × 3.4 mm Silastic cylindrical capsules (each containing 38 mg) and implanted subdermally for up to 5 years' contraception. The exogenous progestin decreases but does not stop secretion of FSH and LH. Progestin-induced thick scanty mucus prevents sperm migration and endometrial development is inhibited. The failure rate is 1 per 100 users per year, but rates are higher in heavier women. Abnormal vaginal bleeding occurs in up to 40% of users within the first year of use, but this improves with time. After removal of the implant, 50% of those wanting to conceive become pregnant within 3 months and 85% within 1 year.

Chapter 32

Genitourinary Tract Disorders

The anatomy and physiology of the urinary tract in the female have been described in the companion volume (*Basic Sciences for Obstetrics and Gynaecology* (Chard and Lilford, 5th edn, 1997)).

Urodynamic Investigations

The simplest tests are a mid-stream urine specimen and an input/output diary. Urinary infection must be excluded before more complex investigations are performed. Other tests include:

1. *Pad test*: in a formal version this consists of weighing a pad before and after a water load (500 ml). The water load is followed by 15 min of rest, then 30 min of gentle exercise, then 15 min of more strenuous exercise.
2. *Uroflowmetry*: urine flow can be measured by a strain transducer, a rotating disk, or a metal strip capacitor. Voided volume should be 150 ml or greater. An average flow rate of less than 15 ml/s is abnormal. Peak flow rate is a valuable parameter.
3. *Cystometry*: this is measurement of pressure and volume during filling and voiding. Pressure (bladder and rectum) is measured with a fluid-filled catheter or a solid pressure transducer at filling rates of 10–100 ml/min. Normal features include a residual volume of less than 50 ml, desire to void at 50–100 ml, capacity greater than 400 ml, and a pressure rise on filling that does not return to the baseline on stopping. Leakage without a rise in detrusor pressure is diagnostic of genuine stress incontinence. Detrusor contractions during filling which the patient cannot suppress indicates detrusor instability.
4. *Videocystourethrography*: this is contrast evaluation of the bladder simultaneous with pressure studies. This is not a routine investigation but may be of value if there is incontinence on standing up, if there is a suspicion of ureteric incompetence, or in the investigation of a neuropathic bladder.
5. *Micturating cystogram*: contrast studies of the bladder without pressure or flow measurements. This may demonstrate abnormalities such as reflux and fistulae.

6. *Urethral pressure measurement*: this is usually measured with a catheter tip transducer. Pressures are measured after filling of the bladder (250 ml) and during withdrawal of the catheter. The urethral closure pressure (difference between urethral pressure and bladder pressure during coughing) is reduced in women with stress incontinence.
7. *Urethral electrical conductance*: flow of current between two electrodes on a catheter detects movement of urine in the urethra. Profiles are measured during withdrawal of the catheter. The test is valuable to confirm stress incontinence, and in the diagnosis and treatment of detrusor instability.
8. *Cystourethroscopy*: this identifies disease of the bladder and urethra but has no specific role in the diagnosis of incontinence.
9. *Ultrasound*: this can be used to estimate residual urine volume and to assess the bladder neck. The latter is a potential alternative to radiology.
10. *Electromyography*: this can be used to assess the pelvic floor muscles. Denervation of these muscles commonly follows delivery and is associated with incontinence.

Stress Incontinence

This is involuntary urine loss on coughing or physical exertion. Urethral sphincter incompetence occurs with descent of the bladder neck and proximal urethra so that pressure in the urethra becomes less than that in the bladder. The fault in the bladder neck may be congenital (e.g. epispadias), or result from childbirth with damage to and denervation of the sphincter, or result from trauma (pelvic fractures, surgery). The condition is further exacerbated by oestrogen deficiency (postmenopause). Symptoms are common in pregnancy and may deteriorate in the puerperium. A cystourethrocele is present in 50% of cases. Frequency, urge incontinence and incontinence on standing are common associated symptoms.

Conservative treatment includes use of absorbent pads, pelvic floor exercises (sometimes aided by vaginal cones), electrical stimulation of pelvic floor muscles, and (rarely) suprapubic catheterisation. Conservative measures are successful in 50% of cases.

Detrusor Instability

Involuntary contraction of the bladder (spontaneous or provoked by coughing, etc.) is a cause of urinary incontinence in up to 10% of all women and 30–50% of women complaining of incontinence. In most cases the cause is unknown. Specific causes include inadequate continence training in infancy, psychosomatic factors, hyperreflexia associated with multiple sclerosis and spinal injury and surgery of the bladder neck. The commonest symptoms are urgency and frequency. Others include nocturia, stress incontinence and coital incontinence. Diagnostic tests have been listed above.

The condition can sometimes be corrected by simple behaviour changes (time of drinking, etc.), or by bladder retraining aided by biofeedback. Drug treatments include anticholinergic agents (propantheline), muscle relaxants (oxybutynin), antidepressants (imipramine), calcium channel blockers (terodiline), antidiuretic hormone analogues (demopressin) and oestrogens. The most effective are oxybutynin (though this may cause dry mouth and an increase in urinary residual volume) and terodiline (side-effects, usually less than oxybutynin, may include hypotension, headache and constipation). The use of transvesical phenol has been described. Surgery is only indicated for the most severe cases and consists of inserting a patch of gut to enhance bladder capacity ("clam" cystoplasty). If detrusor instability is combined with stress incontinence, the latter should be treated first.

Retention of Urine

This may be acute or chronic. Causes include neurological (e.g. multiple sclerosis, cord injury), pharmacological (epidural analgesia), painful lesions of the vulva and perineum (e.g. genital herpes), bladder neck surgery, impacted pelvic masses (e.g. retroverted gravid uterus), foreign bodies in the urethra, ectopic ureterocele, bladder polyps or tumours, cystocele (rarely), prolonged bladder overdistension and psychogenic.

Diagnostic procedures are listed above. Treatment will vary with the cause. For acute cases the first line is catheterisation (urethral or suprapubic). Pharmacological agents may either relax the sphincter (e.g. alpha adrenergic blocking agents such as phenoxybenzamine) or stimulate the detrusor (e.g. muscarinic agents such as bethanechol chloride).

Urinary Tract Infections (UTIs)

UTIs are remarkably common in women. Some 20% of women in the age range 20–65 years suffer at least one attack per year; 50% develop a UTI within their lifetime; and there is a 5% incidence of asymptomatic bacteriuria. The latter is defined as more than 100 000 organisms/ml of mid-stream urine, though women with symptomatic infections may have smaller numbers.

The commonest organisms in UTIs are Gram-negative aerobes from the gut. About 90% of first infections and 70% of reinfections are due to *E. coli*. Other organisms include *Klebsiella* and *Proteus* species, the latter often being associated with renal calculi. The pathogenicity of organisms in the urinary tract depends, among other factors, on the ability of the organisms to adhere to epithelial surfaces via fimbriae on the bacterial cell surface (P fimbriae).

There are many predisposing factors for UTIs, including obstructive lesions, catheterisation, intercourse and the relatively short female urethra. Women of blood groups B or AB are more susceptible to UTI.

Fistulae

Fistulae can enter any part of the genital tract from any viscus; those from the bladder to the vagina are the commonest. The causes include obstetric (pressure necrosis or laceration), surgical or other trauma (including irradiation), inflammatory (genital or bowel including schistosomiasis, lymphogranuloma venereum, tuberculosis, actinomycosis, Crohn's disease (the commonest inflammatory cause), ulcerative colitis and diverticular disease) and neoplastic. Congenital fistulae are rare and include ectopic ureter. In less-developed countries obstetric accidents are the commonest cause; in developed countries it is surgical trauma following abdominal or vaginal gynaecological operations. The main clinical feature is incontinence of urine or faeces. Very minor degrees of incontinence can occasionally be confused with an inflammatory vaginal discharge. The diagnosis is confirmed by exploration with a malleable silver probe, if necessary under general anaesthetic. Other procedures that may help with the diagnosis include contrast radiography (cystography, hysterography, bowel enemas, etc.), local or systemic administration of dyes (methylene blue) and endoscopy. It is important to exclude dual or complex fistulae.

Prevention of many fistulae is possible (better obstetric services and better surgical training). Specific causes (inflammatory, neoplastic) are dealt with as appropriate. Fistulae are usually managed surgically.

Prolapse

Genital prolapse is classified as: descent of the lower third of the anterior vaginal wall (urethrocoele); upper two thirds of the anterior vaginal wall (cystocoele); uterus and cervix; small bowel and omentum (enterocoele); posterior vaginal wall (rectocoele); and vault after hysterectomy. The commonest form is cystourethrocoele. Uterine descent may be first degree (cervix appears at introitus), second degree (cervix (but not uterus) descends through introitus on straining), or third degree (procidentia: uterus entirely outside introitus); the latter may result in ureteric obstruction.

The main predisposing factors to prolapse are damage to the pelvic floor at the time of childbirth and tissue weakening due to oestrogen withdrawal around the menopause. Difficult vaginal delivery (especially after prolonged labour) leads to partial denervation of the pelvic floor muscles, together with damage and stretching of fascia and ligaments. Very rarely, congenital defects such a spina bifida are involved. Black women are less prone to prolapse, as are those who deliver in the squatting position. Obesity, chronic cough, and constipation are other predisposing factors. Symptoms include discomfort or feeling something coming down, urinary frequency, backache, stress incontinence and, in severe cases, urinary retention or difficulty with defaecation. Recurrent urinary tract infection is a common complication of uterovaginal prolapse. The patient is examined in the left lateral position with a Sims' speculum.

Prevention of prolapse includes better obstetric practice, postnatal pelvic floor exercises (Keegels), and hormone replacement therapy. Non-surgical treatment of prolapse includes pelvic floor exercises, electrical stimulation with vaginal electrodes, and pessaries (in the very elderly or very frail patient who is unwilling or unfit for anaesthesia). Pessaries are made of non-toxic plastics (e.g. polyvinyl chloride). The commonest types are the ring (for control of cystocoele and rectocoele) and shelf (for major degrees of prolapse). Pessaries should be changed every year and in postmenopausal women should be combined with the use of an intravaginal oestrogen cream. The main complications are ulceration, with bleeding and sepsis and incarceration.

Surgical treatment of prolapse is discussed in Chapter 35.

Chapter 33

Endometriosis and Adenomyosis

In this condition glands and stroma similar to endometrium are found in the wall of the uterus (adenomyosis) or outside the uterus. Rarely the two conditions coexist.

Endometriosis

Various theories have been put forward on the aetiology of endometriosis. First, retrograde menstruation through the fallopian tubes to the peritoneal cavity is the most popular theory and can explain most cases, but not the occurrence in distant sites such as the pleura. Second, coelomic metaplasia of cells derived from the primitive coelomic epithelium could explain all cases including distant sites. Third, lymphatic/vascular dissemination might explain some rare deep-seated lesions. The fourth theory is mechanical transplantation at the time of surgery.

Endometriosis is associated with infertility, delay in a first pregnancy, and higher social class. There may be a family predisposition but other genetic factors, such as race, do not appear to be important. Antibodies to endometrial antigens have been demonstrated in some patients.

The lesions vary in size from small black dots to large "chocolate" cysts. Local peritonitis leading to dense adhesions is common. Histologically there are glands, stroma, old and new haemorrhage and haemosiderin-laden macrophages. Cyclical changes (e.g. a secretory pattern) may be seen. Malignant change is extremely rare; it is commoner in the ovary than in other sites.

Clinical Features

The highest incidence is in women aged 30–45 years. The commonest presenting features are pain and/or subfertility. The nature of the pain depends on the site, but typically includes pelvic discomfort, backache, deep dyspareunia and secondary dysmenorrhoea. Pain may start just before or at the time of menstruation. There may also be anovulation leading to irregular menstruation and menorrhagia. Rarely, rupture of an endometriotic cyst can lead to an abdominal emergency.

Endometriosis is found in 10–25% of women with gynaecological symptoms and 15% of infertile women. The apparent incidence is increasing because of better diagnosis. About 50% of women with endometriosis may be infertile, the cause may be obstruction due to extensive adhesions, or interference with ovarian or tubal motility, or possibly to release of peritoneal prostaglandins. However, with minor lesions the cause and effect relationship remains contentious.

Examination may reveal hard fixed nodules in the pouch of Douglas, the uterosacral ligaments and the rectovaginal septum. There is often tenderness, especially on movement of the cervix. The uterus may be retroverted and fixed, and the ovaries may be enlarged and immobile. Speculum examination may show lesions in the posterior fornix. In some cases, physical examination may reveal no abormal findings.

The diagnosis is confirmed by laparoscopy, when it can be distinguished from adenomyosis, pelvic inflammatory disease, pelvic cancers, and pelvic "congestion". Sometimes the lesions may be small and difficult to identify; if there is doubt, biopsy should be performed. Other diagnostic methods include measurement of the marker CA-125 (moderately elevated), and imaging by ultrasound or computed tomography or magnetic resonance imaging scanning. Imaging cannot detect small peritoneal lesions.

The severity of the condition is scored according to a system proposed by the American Fertility Society (AFS). However, there is a relatively poor correlation between the score and the clinical features, including infertility.

Endometriosis in Specific Sites

The commonest sites are the ovary and the pelvic peritoneum of the rectovaginal pouch, the uterosacral ligaments and the broad ligament. Other sites that may be involved include (a) the bowel (pelvic colon and rectum, usually confined to the peritoneal and muscular layers; it can lead to rectal pain and bleeding and, rarely, fibrosis and obstruction), (b) the lower genital tract (bluish cystic lesions on the cervix and vagina, rarely in episiotomy scars); (c) the urinary tract (a rare cause of urinary symptoms and sometimes ureteric obstruction); and (d) other sites (umbilicus, laparotomy and caesarean section scars, inguinal region (via the round ligament), limbs and pleura).

Treatment

The main aims of treatment are to relieve pain and to promote fertility if this is an issue.

Earlier attempts to induce a protective state of "pseudopregnancy" with oestrogen–progestagen combinations (e.g. continuous administration of oral contraceptives) have now been replaced largely by use of the isoxazol derivative of 17α-ethinyltestosterone, commonly known as Danazol. This acts at various levels (hypothalamus, pituitary and ovary) to cause anovulation, amenorrhoea and endometrial atrophy. Doses range from 400 mg/day in mild cases up to 800 mg/day for severe lesions, usually continued for 6 months. There is symptomatic improvement in over 85% of cases, and resolution of lesions can be observed in 70–95%. However, the 3-year recurrence rate is as high as 40%.

Side-effects of Danazol may result from gonadotrophin inhibition (reduced breast size, flushes, atrophic vaginitis) or relative androgenisation due to a reduction in sex hormone binding globulin (acne, hirsutes, voice changes). There may also be headaches, fatigue, depression, weight gain, muscle cramps and mild intestinal symptoms. The elevation of low density lipoproteins and reduction of high density lipoproteins and cholesterol reverses rapidly when treatment is stopped. If anovulation continues to be a problem after a course of Danazol, then there is no contraindication to ovulation-inducing drugs.

Other medical treatments include gestrinone, a synthetic 19-norsteroid that has effects and side-effects generally similar to those of Danazol, and analogues of GnRH. Substitution of amino acids gives these a prolonged half-life; they can be administered intranasally (nafarelin, buserelin) or by depot injection (goserelin). Continuous administration downregulates the pituitary and produces a "medical castration". This yields rapid symptomatic relief and some resolution of deposits. The efficacy is similar to that of Danazol. The side-effects are the result of the hypo-oestrogenic state – flushes, headaches, vaginal dryness, loss of libido, osteoporosis.

Surgical treatment involves removal of ovarian tissue and/or endometriotic deposits via laparotomy or laparoscopy. In a younger woman who may wish to be fertile, the procedure includes division of adhesions, removal or diathermy (electro or laser) of endometriomas and implants, mobilisation of the ovaries and tubes, and sometimes salpingostomy. In older women a more radical approach may be taken, with total hysterectomy and bilateral salpingo-oophorectomy. The procedures may be easier after a short preoperative course of Danazol and GnRH analogue.

The choice of treatment according to severity is as follows:

1. *Mild endometriosis*: if the symptoms are not especially severe, and fertility is required, then simple observation is adequate. If symptoms are more severe, a course of Danazol should be given and stopped as soon as pregnancy is planned;
2. *Moderately severe endometriosis* (AFS III or more): a course of Danazol should be tried but, if not successful, surgery is necessary;
3. *Severe disease*: this will almost always require surgery.

Adenomyosis

Ingrowths of endometrium are seen in the uterine muscle. It is associated with repeated pregnancies (unlike endometriosis, which occurs in women of low parity), or overvigorous curettage. The diagnosis can be made only histologically.

The uterus is usually symmetrically enlarged, though sometimes there are multiple, non-encapsulated localised lesions. The cut uterus shows pale areas with central "blood spots"; occasionally there are blood-filled cystic spaces. Histologically there are glandular cells and stroma surrounded by muscle cells. The tissue responds to hormones and may show secretory changes, a decidual reaction during pregnancy, or sometimes cystic glandular hyperplasia.

The condition is seen in middle-aged multiparous women, often from higher socioeconomic groups. The principal features are menorrhagia, secondary dysmenorrhoea and sometimes dyspareunia. The symptoms are similar to those of fibroids and dysfunctional uterine bleeding. Unless more children are wanted the treatment is total hysterectomy with conservation of ovaries.

Chapter 34
Congenital Uterine and Vaginal Abnormalities

An absent or rudimentary uterus is incompatible with reproductive function. Patients present with primary amenorrhoea. There is no treatment.

If a blind vagina is present (note separate embryological development from urogenital sinus in addition to müllerian component), androgen insensitivity must be considered. Coincident renal tract abnormalities should be excluded by means of ultrasound and intravenous pyelogram studies. Surgical reconstruction of the vagina is possible, though intermittent pressure with dilators may give equally good results. Amenorrhoea together with abdominal pain, and difficulty with micturition, should suggest haematocolpos (the accumulation of menstrual blood retained by an imperforate lower vaginal membrane). Surgical incision generally suffices. A vaginal septum may cause dyspareunia, or it may tear in labour; in both cases it should be removed.

The uterus, cervix and fallopian tubes are derived from the fused müllerian ducts. Minor fusion defects (arcuate and bicornuate uterus) are common (1% of the general female population) and usually symptomless. Less common are septate and subseptate uterus, as well as complete duplication of the uterus and cervix (uterus didelphys). Though rarely symptomatic, these latter may present with repeated spontaneous abortion (congenital anomalies account for 10% of all cases of recurrent miscarriage), repeated transverse lie or obstructed descent of the head in late pregnancy or labour. These conditions may now be treated by hysteroscopic ablation of the septum rather than open laparotomy procedures such as Strassman's operation. Rupture of a pregnancy-containing rudimentary horn is symptomatically similar to ruptured ectopic pregnancy and can be lethal.

Chapter 35

Gynaecological Operations

Vulvectomy

Vulvectomy may be simple or radical. The former may be total or partial depending on the extent of disease. The major indications for simple vulvectomy are extensive vulvar intraepithelial neoplasia, intractable pruritus vulvae, Paget's disease or after failed medical treatment of vulval dystrophies.

In total simple vulvectomy, the skin incision should be elliptical to include the clitoris, the labia and the fourchette; the inner incision extends elliptically just above the urethral opening down to the vaginal mucocutaneous junction. The wound is closed symmetrically. A Foley catheter is maintained for a few days to keep the vulva dry.

Radical vulvectomy is usually combined with bilateral dissection of the groin nodes in the surgical treatment of carcinoma of the vulva. The skin incision is generally butterfly shaped. Most surgeons would also remove the distal centimetre of the urethra.

Vulvectomy may be followed by wound breakdown due to infection, secondary haemorrhage or haematoma. Late complications include thromboembolism, lymphocysts, oedema of the legs and vaginal prolapse.

Vaginal Repair

Anterior colporrhaphy is indicated for repair of a cystocele, posterior colpoperineorrhaphy for that of a rectocele, while the Manchester (Fothergill) repair (approximating shortened cardinal ligaments anterior to the amputated cervix and anterior repair) is indicated for all degrees of prolapse. When the cystocele is accompanied by urinary stress incontinence, both are effectively treated by a Burch colposuspension. Prolapse of the vaginal vault (enterocele) is best treated by vaginal excision and ligation of the hernial sac and approximation of the uterosacral ligament and endopelvic fascia anterior to the rectum. Recurrent prolapse (or failed repair) occurs in up to 25% of women.

Colposcopy and Cervical Operations

The indications for colposcopy include abnormal cervical cytology (CIN III or recurrent CIN II) and suspicious vaginal or cervical lesions even if cytology is negative. Acetic acid (5%) turns the abnormal epithelium white. The features of cervical intraepithelial neoplasia include mosaicism and punctation. Vascular changes, including atypical vessels, are seen in invasive disease.

Cone biopsy is indicated for the evaluation of abnormal cytology when colposcopy is unavailable, if the whole of the transformation zone is not visualised or when there is any suspicion of an invasive or endocervical lesion. The carbon dioxide laser may be used instead of the knife. The main complications are haemorrhage (5–20%), cervical stenosis (3–30%), incomplete excision with the risk of invasive cancer and increased pregnancy complications, e.g. preterm delivery and mid-trimester miscarriage.

When the entire lesion is seen colposcopically and biopsy has excluded invasive cancer or an endocervical abnormality, local destructive therapy of the transformation zone may be achieved by (a) electrodiathermy at a depth of up to 1.5 cm, usually under general anaestheisa (large-loop excision of transformation zone); (b) cold coagulation (50–120 °C) at a depth of 3–4 mm; (c) cryocautery with a 3 min freeze thaw cycle at a depth of 3–4 mm; (d) laser destruction (carbon dioxide) at a depth of between 5 and 10 mm. The last three are usually performed as outpatient procedures without anaesthesia, but only electrodiathermy and laser vaporisation are possible under colposcopic vision. The overall success rate in the treatment of cervical intraepithelial neoplasia is approximately 95%. Pain, discharge and bleeding are complications seen to a greater or lesser extent following all local ablative treatment methods.

Dilatation and Curettage

Dilatation and curettage (D&C) is the most frequently performed operation in gynaecology. Pelvic examination precedes insertion of the uterine sound. After careful dilatation to 7–10 mm, the uterine cavity is carefully explored and systematically curetted. Complications include: (a) laceration of the cervix; (b) primary or secondary haemorrhage; (c) perforation of the uterus with or without peritonitis, peritonism, bowel damage or haemorrhage into the broad ligament; (d) pelvic cellulitis and parametritis; and (e) Asherman's syndrome (intrauterine adhesions).

Hysterectomy

Hysterectomy may be total (simple), partial (subtotal) or radical. The former may be performed abdominally or vaginally for the management of benign uterine disease. Abdominal hysterectomy is also used to treat stage I grade I uterine carcinoma and as part of the management of ovarian carcinoma. The abdominal procedure includes division of the round and infundibulopelvic ligaments, incision of the uterovesical peritoneum to separate the bladder followed by division of the uterine vessels, uterosacral ligaments and vaginal angles; some or all of these may be performed by the endoscopic route. The pedicles are ligated and the vaginal edges sutured. Haemorrhage, infection and vault haematoma are the most common complications. Ureteric damage with subsequent fistula formation is always possible.

The main indication for vaginal hysterectomy is genital prolapse, but it can also be used to treat benign uterine disease provided the uterus is no larger than a 12-week pregnancy and there is no evidence of endometriosis or pelvic inflammatory disease, which may restrict access. A circumferential incision around the cervix allows the vaginal skin to be reflected. The cervicovesical ligament is divided and the pouch of Douglas opened. The supporting uterine ligaments are divided and the uterovesical pouch opened. Finally the uterine vessels and tuboovarian pedicles are divided. The vaginal vault is closed and supported using the pedicles. The most common complication is a vaginal vault haematoma.

Radical abdominal (Wertheim's) hysterectomy is combined with pelvic node dissection in the treatment of carcinoma of the cervix. The procedure includes removal of the uterus, at least the upper one third of the vagina and the parametrium out to the pelvic side wall. All visible pelvic nodes from the bifurcation of the aorta down to the cut end of the uterine artery are removed. Intraoperative complications include haemorrhage, injury to the urinary or intestinal tract and nerve trauma. Postoperative infection, ileus and thromboembolic disease may occur as well as the formation of lymphocysts and fistulae (vascular, ureteric, bladder, bowel).

Tubal Surgery

Tubal surgery is best performed using an operating microscope and microsurgical technique with irrigation, careful haemostasis, avoidance of peritoneal damage and the use of fine, non-absorbable sutures. Adhesions can be removed (salpingolysis), fimbriae released (fimbrioplasty) or a new tubal stoma created (salpingostomy). In many cases these procedures can be done by videolaparoscopic techniques. Tubotubal or tubocornual anastomosis is also possible, as is reimplantation of the tube to the cornua. Some give steroids and antibiotics postoperatively whereas others perform peritoneal lavage with Ringer's lactate solution and/or Dextran 70 in an attempt to reduce adhesions. Subsequent ectopic pregnancy is the major complication with all types of surgery. Ectopic pregnancy can be treated by linear salpingostomy, or partial or total salpingectomy at conventional laparotomy or by videolaparoscopic techniques. Topical or systemic administration of medications such as methotrexate are appropriate in some women.

Ovarian Surgery

Ovarian cystectomy involves enucleating a cyst while retaining the major part of the ovary. Most capsulated benign ovarian cysts can be removed in this way provided some ovarian tissue is retained. These and other ovarian procedures may be performed at laparotomy or by videolaparoscopic techniques.

Removal of a diseased ovary (ovariotomy) may form part of the total management of carcinoma of the ovary (together with total hysterectomy and omentectomy). Very large ovarian masses may distort the pelvic anatomy and place the ureter at risk of injury during their removal. Adhesions increase the possibility of bowel damage.

Laparoscopy

Laparoscopy may be used diagnostically and/or therapeutically. The majority of gynaecological operations can be performed by videolaparoscopy. A pneumoperitoneum is produced by insufflating 2–4 litres of carbon dioxide through a Verres needle inserted subumbilically. The laparoscope is inserted via the trochar. Other instruments, typically three, are inserted paraumbilically ($\times 2$) and suprapublically into the abdominal cavity to improve organ handling and to perform endoscopic surgery (e.g. haemostasis, ovarian cysts, endometriosis, extraperitoneal lymph node dissection, colposuspension, adhesiolysis, linear salpingostomy for ectopic pregnancy, neosalpingostomy and laparoscopically assisted hysterectomy) or assisted conception techniques (e.g. egg collection for IVF and GIFT). Complications (2%) include surgical emphysema, vascular, bowel and urinary tract injury or the need to proceed to open operation.

Hysteroscopy

Diagnostic hysteroscopy can be performed without cervical dilatation and under local anaesthesia as an outpatient or office procedure. Operative hysteroscopy for myomectomy, polypectomy, division of adhesions and endometrial ablation is usually performed under general anaesthesia. Endometrial resection may be more easily performed after ovarian suppression with Danazol or GnRH analogues. Potential immediate risks include uterine perforation, infection, bleeding and water intoxication (if the procedure is prolonged). Long-term complications include haematometra (2%) and failure to cure menorrhagia (15%). Simultaneous laparoscopy guards against the consequences of uterine perforation.

Pre- and Postoperative Care

Women with pre-existing medical diseases, particularly if they are taking medication, require careful assessment. Ideally, obesity, smoking and anaemia are corrected preoperatively.

Postoperative complications may be: (a) local, involving the operation site itself, e.g. reactionary haemorrhage, wound haematoma and dehiscence, vault haematoma, pelvic infection and fistula formation; or (b) general, e.g. pulmonary collapse, infection, embolism, urinary retention or infection and deep vein thrombosis. These complications may be immediate (within the first 24 h), early (within the first 2–3 weeks) or late (any time thereafter) in relation to the operation.

Subject Index

Abdominal pain 93-5
Abdominal pregnancy 25
Abdominal wall anomalies 41
Abnormalities of implantation 16
Abortion, induced 20-1
Abruptio placentae 53-4
Acute renal failure 80
Addison's disease 92
Adenomyosis 205-6
Adrenal disease 92-3
Adult respiratory distress syndrome 73-4
Alpha thalassaemia 101
Amenorrhoea
 primary 144-5
 secondary 146
Amniocentesis 133
Amniotic fluid
 disorders involving 57-8
 embolism 4, 58
 volume 13, 57
Amniotic Fluid Index (AFI) 57
Anaemia 99
 neonatal 131
Anaesthesia
 epidural 109-10
 in maternal mortality 4
Analgesia 109-10
Androgen insensitivity syndrome 144
Antenatal care 11-14
Antenatal management, twins 65
Antepartum cardiotocography (CTG) 13
Antepartum haemorrhage (APH) 53-5
 "indeterminate" 55
 marginal 55
Apgar scoring table 125
Aplastic anaemia 103
Appendicitis 94
Asphyxia 125-7
 causes and effects 126
 clinical features 127
 definition 125
 incidence 125
 outcomes 127
Asthma 70-2
 management in pregnancy 71

management of labour 71
Autoimmune disease 16
Autoimmune haemolytic anaemia 103
Autosomal dominant inheritance 33
Autosomal recessive inheritance 33

Barrier methods of contraception 194
Behcet's syndrome 156
Bell's palsy 76
Benign intracranial hypertension 75
Beta thalassaemia 101
Biophysical profile 14
Biparietal diameter (BPD) 8
Birth rate 3
Bleeding
 postmenopausal 152
 see also Haemorrhage
Bleeding disorders, congenital 104
Blood pressure 47
Breast abscess 123
Breast cancer 96
Breast development 123
Breast feeding 123
Breech presentation 113-15
 version for 135
Brenner tumours 186
Brow presentation 113

Caesarean section 66, 122, 136, 173
Calendar method of contraception 193
Candida albicans 158, 196
Candida vaginitis 158
Cardiac disease 68-9
Cardiovascular disease 67-70, 151
Cardiovascular system, physiological changes 67
Carpal tunnel syndrome 76
Cephalopelvic disproportion (CPD) 108, 111-12
Cerebrovascular disease 75
Cervical cap 194
Cervical incompetence 16, 18
Cervical intraepithelial neoplasia (CIN) 169-70
Cervical suture 135
Cervicitis 159

Cervix
　benign tumours 169
　carcinoma 171–3
　carcinoma-in-situ 169–70
　microinvasive carcinoma 170–1
Chest infections 121
Chlamydia 159
Chlamydia trachomatis 155, 160
Chlamydial infection 160
Choriocarcinoma 27, 29–30, 185
　clinical features 29
　diagnosis 29
　treatment 29–30
Chorionic villus sampling (CVS) 133
Chromosomal anomalies 41, 64
Chromosomal disorders 31
Chromosome number, abnormalities 31–2
Chromosome structure, abnormalities 32–3
Chronic pelvic inflammatory disease 160–1
Chronic renal disease 77–9
　clinical features 78
　pregnancy management 79
Climacteric 149
　anatomical changes 150
　pathology 150–1
　stages 149
　see also Menopause
CNS abnormalities 40
Coagulation disorders 103–5
Coeliac disease 84
Coitus interuptus 193
Colposcopy 210
Combined oral contraceptive pill (COCP) 196–7
Condoms 194
Congenital abnormalities 31–42
　screening 42
　uterus 207
　vagina 207
Congenital adrenal hyperplasia 144–5
Congenital anomalies, uterus 18–19
Congenital bleeding disorders 104
Constipation 83–4
Contraception 193–8
Cord prolapse 117–18
Cordocentesis 134
Corticosteroids in premature labour 60
Crohn's disease 201
Crown-rump length (CRL) 8
Cushing's syndrome 92, 154
Cystic fibrosis 73
Cystitis 121
Cytology 210

Death. *See* Mortality
Delivery
　forceps 136–7
　normal 108
　ventouse 137
Depression 121

Dermoid cysts 183
Detrusor instability 200–1
Diabetes 85–90
　classification 88
　gestational 88–90
　management of labour 88
Diabetes insipidus 93
Diaphragms 194
Diarrhoea 84
Dilatation and curettage (D&C) 210
Disgerminoma 185
Disseminated intravascular coagulation (DIC) 55
Dizygotic (DZ) twinning 63
Donovania granulomatosis 155
Doppler blood flow 14
Dual energy X-ray absorptiometry (DEXA) 151
Dysfunctional uterine bleeding (DUB) 142–4, 149
　pathology 143
　treatment 143–4
Dysmenorrhoea 147
Dyspareunia 148

Eclampsia 50–1
Ectopic pregnancy 23–5
　causes 23
　clinical features 24
　treatment 24
Electronic fetal heart rate monitoring (EFM) 117
Endocarditis prophylaxis 67
Endocrine deficiencies 16
Endocrine diseases 85–93
Endodermal sinus tumour 185
Endometriosis 203–5
　aetiology 203
　clinical features 203–5
　specific sites 204
　treatment 204–5
Endometrium
　benign tumours 175
　carcinoma 175–7
　stromal tumours 181
Environmentally-induced abnormalities 34, 36–7
Epidural anaesthesia 109–10
Epilepsy 74–6
Episiotomy and repair 137
Erb's palsy 118
Extra-corporal membrane oxygenation (ECMO) 126

Face presentation 113
Female condom 194
Fertility awareness 193
Fetal abnormality 15–16
Fetal breathing movements (FBM) 13

Fetal growth, ultrasound measurement 12–13
Fetal heart rate (FHR) 13, 116, 126
Fetal malformations 64
Fetal mortality 5
Fetal movements 12
Fetal well-being
  monitoring 115–17
  tests 12
Fetoscopy 134
Fetus in labour 108
Fibroids 179–80
  in pregnancy 180
  metastasising 180
Fibroma 183
Fistulae 201–2
Fluid retention, management 51
Forceps delivery 136–7

Gardnerella vaginitis 158
Gastrointestinal anomalies 41
Gastrointestinal disease 83–5
Gastrointestinal tract, changes in pregnancy 83
Gene defects, single 33
Genetic defects 31
  diagnosis 34
Genital infections 155–62
Genital prolapse 202
Genital tuberculosis 161
Genital warts 157
Genitourinary disease 76–80
Genitourinary tract disorders 199–202
Germ cell tumour 185
Gestational age
  assessment 7–8
  ultrasound parameters 8–9
Gestational diabetes 88–90
Gestational hypertension 48
Gestational trophoblastic tumours 27–30
GIFT 191
Glucose homeostasis 85
Glucose tolerance test 90
Gonadotrophin, chorionic (hCG) 28–9
Gonococcal infection 160
Gonococcal salpingitis 160
Gonococcal vaginitis 158
Granuloma inguinale 155
Granulosa cell tumours 184
Growth retardation 64
Guillain-Barré syndrome 76

Haematological disorders 99–105
  neonate 131
Haematological malignancies 105
Haemodialysis 79
Haemoglobinopathies 100–2
Haemolytic anaemia 102–3
Haemophilia A 104

Haemorrhage
  antepartum. See Antepartum haemorrhage (APH)
  in maternal mortality 4
  postpartum 66, 119–20
Haemorrhagic disease, neonatal 132
Haemorrhoids 84
Haemostatic disorders 103
hCG levels 28–9
Heart disease 67–70
  maternal and fetal prognosis 70
  see also Cardiovascular
Heartburn 83
Hepatitis 161–2
Hermaphroditism 145
Herpes gestationis 97
Herpes simplex virus (HSV) infection 156
High frequency oscillatory ventilation (HFOV) 126
Hirsutism 153–4
  adrenal causes 154
  ovarian causes 153–4
  treatment 154
HIV infection 161
Homozygous Hb C disease 102
Hormone contraception 196–8
Hormone replacement therapy (HRT) 152, 193
Human immunodeficiency virus (HIV), pulmonary complications 72
Hydatidiform mole 27–8
  aetiology and distribution 27–8
  treatment 28
Hydramnios 57
Hyperbilirubinaemia 128–30
Hyperemesis gravidarum 9
Hypergonadotrophic hypogonadism 145
Hyperprolactinaemia 146
Hypertensive disorders 47–51, 87
  classification 47–8
  gestational 48
  in maternal mortality 4
  pregnancy-induced 48
  prognosis 51
  see also Eclampsia; Pre-eclampsia
Hyperthyroidism 91
Hypoparathyroidism 82
Hypopituitarism 93
Hypothermia, neonatal 130
Hypothyroidism 91
Hypoxic-ischaemic encephalopathy (HIE) 126–7
Hysterectomy 210–11
Hysteroscopy 212

Idiopathic respiratory distress syndrome (RDS) 128
Immunological factors 16
Impacted shoulder 118

Impetigo herpetiformis 98
Induction of labour 109
Infection
  chlamydial 160
  genital 155-62
  gonococcal 160
  in miscarriage 16, 19
  neonate 130-1
  pelvic 120-1, 159-60
  puerperium 120-1
  respiratory tract 72-4
  urinary tract 77, 201
  vagina 157-9
  vulva 155
Infective diarrhoea 84
Infertility 189-91
  aetiology 189
  definition 189
  investigations 190
  management 189
  treatment 190-1
Inflammatory bowel disease 85
Injectable steroids 197-8
Insulin secretion 85
Intestinal obstruction 94-5
Intra-amniotic infection 58
Intrauterine adhesions 16
Intrauterine contraceptive device (IUCD) 195
Iron deficiency anaemia 99-103

Jaundice 85, 128-30
  causes, risk factors and clinical features 129

Kyphoscoliosis 73

Labour
  abnormal 111-18
  delay in second stage 117
  diagnosis 107, 111
  induction 109
  mechanisms and course 107
  monitoring of fetal well-being 115-17
  normal 107-10
  normal delivery 108
  prolonged 111-12
  stages 108
  see also Delivery; Premature labour
Labour management
  in premature labour 60
  twins 65
Laparoscopy 212
Leiomyoma 179, 180
Leiomyosarcomas 180
Limb deformities 42
Lipoproteins 151
Liver disease, clinical features 86-7

Lung. See Respiratory system
Lymphogranuloma venereum 155

Malignant disease 95-6
  clinical features 95
Malignant epithelial tumours 186-8
Malignant germ cell tumours 185
Malignant teratoma 185
Malpresentations 112-15
Marfan's syndrome 83
Masculinising tumours 184-5
Maternal diseases 16, 67-105
Maternal mortality 4-5, 122
Maternal weight gain 12
Meconium stained liquor (MSL) 115-16
Megaloblastic anaemia 100
Mendelian inheritance 33
Mendelson's syndrome in maternal mortality 4
Menopause 149-52
  clinical features 151-2
  diagnosis 151
  premature 147
  see also Climacteric
Menorrhagia 141-2
  causes 142
  diagnosis 141
Menstrual disorders 141-8
Menstruation, mechanisms 142-4
Metabolic diseases 82-3
Migraine 75
Miscarriage 15-21
  causes 15-17
  definition 15
  incomplete 17
  inevitable 17
  missed 17-18
  recurrent 19
  septic 19
  threatened 17
  types 15
Monitoring, fetal well-being 115-17
Monosomy XO 32
Monozygotic (MZ) twinning 63
Mortality
  fetal 5
  maternal 4-5, 122
  neonatal 5
  perinatal 65
    in premature labour 61
  postneonatal 5
Mucinous cystadenoma 183
Multiple pregnancy 16, 63-6
  complications 64-5
Multiple sclerosis 75
Multisystem abnormalities 34, 38-9
Musculoskeletal disease 80-3
Mycoplasma hominis 160
Myometrium, benign tumours 179-80

Subject Index

Natural family planning 193
Neisseria gonorrhoea 158
Neonatal hypothermia 130
Neonatal mortality 5
Neonatal polycythaemia 131
Neonate 125–32
  examination 125
  haematological disorders 131
  haemorrhagic disease 132
  infection 130–1
  resuscitation 127–8
  specific problems 128–32
Neurological disease 74–6
  clinical features 75
Neuromuscular disorders 75
Neuropathies 76
Norplant system 198
Nutritional disorders 75

Oblique lie 115
Obstetric operations 133–8
Occipitoposterior position 112–13
Oligohydramios 57–8
Osteomalacia 82
Osteoporosis 150
Ovarian cancer 96
  factors affecting incidence/aetiology 186
  staging 187
Ovarian cystectomy 211
Ovarian surgery 211
Ovary
  benign tumours 183–8
  factors affecting incidence/aetiology of ovarian cancer 186
  five year survival rates for ovarian cancer 188
  histological classification of tumours 187
  malignant epithelial tumours 186–8
  malignant germ cell tumours 185
  special tumours 184

Pancreatitis 94
Papular dermatitis 97
Paraplegia 75
Pearl Index 193
Pelvis
  in labour 108
  infection 120–1, 159–60
  size and shape 108
Peptic ulcer disease 84
Perinatal morbidity, in premature labour 61
Perinatal mortality 65
  in premature labour 61
Perinatal mortality rate (PNMR) 5
Persona system of contraception 194
Phaeochromocytoma 93

Phenylketonuria 82–3
Pituitary disease 93
Pituitary tumours 93, 146–7
Placenta, manual removal 138
Placenta praevia 54–5
Placental function tests 12
Plaques of pregnancy 97
Poisons in miscarriage 16
Polycystic ovarian disease 146
Polycythaemia, neonatal 131
Polyhydramnios 64
Postcoital contraception 197
Postmenopausal bleeding 152
Postmenopausal endocrinology 149–50
Postneonatal mortality 5
Postoperative care 212
Postpartum haemorrhage 66, 119–20
Postpartum thyroiditis 92
Precocious puberty 141
Pre-eclampsia 48–50, 64
  coagulation system in 49
  liver involvement 49
  maternal arterial system in 49
  renal system in 49
  treatment 50
Pregnancy
  diagnosis 7
  termination of 134
Premature labour 59–61
  definition 59
  management 59–60
  outcome 60–1
  prevention 59
  tocolytic therapy 59–60
Premenstrual syndrome 147–8
Prenatal diagnosis 34
  invasive procedures 35
Preoperative care 212
Presentation
  abnormal 112–15
  compound 115
  normal 108
Preterm birth 64
Preterm labour 65
Preterm premature rupture of membranes 61
Primary hyperparathyroidism 82
Progesterone-only pill (POP) 197
Prolapse
  cord 117–18
  genital 202
Prurigo gestationis 97
Pruritic urticarial papules 97
Pruritus gravidarum 97
Pruritus vulvae 155
Psychiatric disorders, puerperium 121–2
Puberty 141
Puerperium 119–23
  definition 119
  infection 120–1
  psychiatric disorders 121–2

Puerperium (continued)
   thromboembolism 122-3
   urinary tract disorders 121
Pulmonary embolism 122
Pyelonephritis 121

Quadruplets 63

Renal tract
   abnormalities 41-2
   functional changes 77
Renal transplantation 79
Respiratory drugs 71-2
Respiratory system 70-4
Respiratory tract infections 72-4
Resuscitation of the newborn 127-8
Rhesus antigen, types 45
Rhesus disease 45-6
   diagnosis 45
   management 46
   prognosis 46
Rhesus incompatibility 16
Rheumatoid arthritis 81
Rhythm method of contraception 193
Rickets 82
Rubella 34
Rupture of membranes, preterm premature 61
Ruptured uterus 118

Sarcoidosis 73
Scleroderma 81
Screening, congenital abnormalities 42
Serous cystadenoma 183
Sexual maturation 145
Sexually transmitted diseases (STDs) 159, 161-2
SHBG 154
Shoulder, impacted 118
Sickle cell anaemia 102
Skeletal dysplasias 42
   see also Musculoskeletal disease
Skin disease 96-8
Skin disorders 98
Spermicides 194
Sterilisation 134-5
Steroids, injectable 197-8
Stillbirth 5
Stress incontinence 200
Sudden infant death syndrome (SIDS) 6
Syphilis 34
Systemic lupus erythematosus (SLE) 80-1

Termination of pregnancy 134
Thalassaemia 101
Theca cell tumours 184

Thrombocytopenia 103-4
Thromboembolism
   in maternal mortality 4
   puerperium 122-3
Thrombosis in maternal mortality 4
Thrombotic disorders 104-5
Thyroid disease 90-2
Thyroid function
   in pregnancy 90-1
   physiological changes 90
   tests 91
Transverse lie 115
Trichomonas vaginalis 157
Triplets 63, 64, 66
Trisomy 13 32
Trisomy 18 32
Trisomy 21 32
Tubal surgery 211
Tuberculosis 72-3
   genital 161
Tumours. See specific histological types
Turner's syndrome 145
Twin pregnancy 63-6
Twinning mechanisms 63-4
Twins
   antenatal management 65
   labour management 65

Ulceration, vulva 155-7
Undermasculinised male 145
Ureaplasma urealyticum 160
Ureteric obstruction 94
Urinary tract disorders, puerperium 121
Urinary tract infections (UTIs) 77, 201
Urine retention 201
Urodynamic investigations 199-200
Uterine abnormalities 16
Uterine activity 107-8
Uterine bleeding, abnormal 143
Uterine corpus, carcinoma 176
Uterus
   congenital abnormalities 18-19, 207
   malignant non-epithelial tumours 180-1
   ruptured 118

Vagina
   benign tumours 167
   carcinoma 167-8
   congenital abnormalities 207
   infection 157-9
   repair 209
Vaginitis 157-9
Vascular disease 151
Vascular system. See Cardiovascular system; Cerebrovascular system
Vasectomy 135
Vault cap 194
Ventouse delivery 137

Version for breech presentation 135
Videolaparoscopy 212
Viral hepatitis 161-2
Virilism 153
Von Willebrand's disease 104
Vulva
  benign swellings 163
  benign tumours 163
  carcinoma 165-6
  infections 155
  premalignant conditions 164-5
  rare malignant tumours 166
  ulceration 155-7

Vulval dystrophies 164
Vulvectomy 209

Warts, genital 157
Wertheim's hysterectomy 173

X-linked recessive inheritance 33

Yolk sac tumour 185
Yuzpe regimen 197